THE DEMOCRACY OF SUFFERING

THE DEMOCRACY OF SUFFERING

LIFE ON THE EDGE OF CATASTROPHE, PHILOSOPHY IN THE ANTHROPOCENE

Todd Dufresne

McGill-Queen's University Press

Montreal & Kingston • London • Chicago

ISBN 978-0-7735-5875-5 (cloth)
ISBN 978-0-7735-5876-2 (paper)
ISBN 978-0-7735-5961-5 (ePDF)
ISBN 978-0-7735-5962-2 (ePUB)

Legal deposit third quarter 2019
Bibliothèque nationale du Québec

Printed in Canada on acid-free paper that is 100% ancient forest free (100% post-consumer recycled), processed chlorine free

This book has been published with the help of a grant from the Canadian Federation for the Humanities and Social Sciences, through the Awards to Scholarly Publications Program, using funds provided by the Social Sciences and Humanities Research Council of Canada.

We acknowledge the support of the Canada Council for the Arts.
Nous remercions le Conseil des arts du Canada de son soutien.

Library and Archives Canada Cataloguing in Publication

Title: The democracy of suffering : life on the edge of catastrophe, philosophy in the Anthropocene / Todd Dufresne.

Names: Dufresne, Todd, 1966- author.

Description: Includes bibliographical references and index.

Identifiers: Canadiana (print) 2019012010X | Canadiana (ebook) 20190120177 | ISBN 9780773558762 (softcover) | ISBN 9780773558755 (hardcover) | ISBN 9780773559615 (ePDF) | ISBN 9780773559622 (ePUB)

Subjects: LCSH: Climatic changes. | LCSH: Nature—Effect of human beings on. | LCSH: Reason. | LCSH:

Capitalism. | LCSH: History, Modern.

Classification: LCC GE149 .D84 2019 | DDC 304.2/5—dc23

No more than one or a few decades remain before the chance to avert the threats we now confront will be lost and the prospects for humanity immeasurably diminished.

"World Scientists' Warning to Humanity (1992)"
signed by 1,700 scientists in 1992

To prevent widespread misery and catastrophic biodiversity loss, humanity must practice a more environmentally sustainable alternative to business as usual. This prescription was well articulated by the world's leading scientists 25 years ago, but in most respects, we have not heeded their warning. Soon it will be too late to shift course away from our failing trajectory, and time is running out. We must recognize, in our day-to-day lives and in our governing institutions, that Earth with all its life is our only home.

"World Scientists' Warning to Humanity: A Second Notice (2017)"
signed by more than 20,000 scientists in 2018

CONTENTS

ACKNOWLEDGMENTS

This book benefited from conversations I have had with lots of people, including students too numerous to mention. It also benefited from academic "friends" on Facebook, who often make smart posts, and from the constant flow of content on the environmental, political, and economic fronts. Kant engaged with his everyday in a Berlin newspaper; today we engage with the everyday through an incredible flow of facts and opinions, a "news feed," that exists at our fingertips. I readily admit that, without this source, the resulting book would have been different – probably very different.

During the research and writing of this book (roughly 2011–18) I taught a few courses of relevance to the ideas expressed herein. I thank my philosophy colleagues for indulging my interests and the many students who did the same. It was invaluable. Imre Szeman invited me to contribute to *Fueling Culture: Energy, History, Politics* (2017), for which I wrote a brief chapter called "Future." That chapter inspired the direction and structure of this book, which already was well underway, so I owe him for that opportunity. I would also like to acknowledge friends and colleagues who read parts or whole sections of earlier versions of this manuscript and made suggestions or provided commentary. Richard Maundrell read an early draft, generously making a detailed list of helpful impressions, criticisms, and suggestions. Clara Sacchetti read a very early and a much later iteration of the book and shared her impressions many times. Christine Daigle read an early draft of the complete book, shared criticisms, and then visited Lakehead University where she delivered a lecture on posthumanism. Justin Currie, a philosophy student, read a version of the book and provided a list of queries, remarks, and suggestions. My nephew Matthew Dufresne read it and shared his impressions between sets. Greg Deschodt read a passage about the dystopic future and shared his views. And in the summer of 2018 Daniel Keyes read the first third of the manuscript and shared his thoughts. I am honoured by these interactions and am pleased to thank each person by name.

Georgina Chuatico helped me with some of the basic research for this book, a position supported by a grant made available by the dean of Social Sciences and Humanities. My friend Mark Nisenholt tweaked many of the public domain images used in the design of the book. And five artists generously allowed me to reproduce, without any qualms, their works in black and white: Nisenholt, Andy Singer, Maria Whiteman, Michael Dal Cerro, and Dwayne Booth (a.k.a. Mr Fish). Thank you very much. Kathy Kaszas extended her usual goodwill. Martha Webb gave me advice about the manuscript. And so, too, did Mark Abley, my very fine editor at MQUP; Shelagh Plunkett, my thoughtful copy editor; and two anonymous reviewers. They all helped me revise and improve my remarks – one well-informed reviewer generously repairing gaps in my understanding of the very diverse, very broad literatures on the Anthropocene. Thanks.

I dedicate this book to Clara and Chloe. To the moon and back.

PREFACE, OR WHAT'S LOVE (OF WISDOM) GOT TO DO WITH CLIMATE CHANGE?

Philosophers have always attempted to account for how it is that human beings can *know* the external world. By the time of the Enlightenment, they had come up with four basic theories. One, we could follow René Descartes and insist that the foundation for knowledge of the world is given *a priori* – certain because known intellectually, like mathematics, before or outside of sense experience. This counter-intuitive perspective is called rationalism. Two, we could follow John Locke and insist that the foundation for know-ledge is *a posteriori* – certain because derived from sense experience of the external world. This reassuring perspective is called empiricism. Three, we could agree with David Hume that empirical knowledge of the world is only a probable (not certain) knowledge, and that probable knowledge is the best we can hope for. This cautious perspective remains a restricted form of empiricism. Or four, we could drop both theories of rationalism and empiricism and devise a new theory that better accords with our con-viction that our knowledge of the external world is certain and not merely conceptual, experiential, or probable.

Immanuel Kant famously opted for the fourth and most difficult choice, thereby transforming the way we think about how human beings come to know the external world. Cue his "Critical Philosophy," a brilliant synthesis of what is right and useful about both rationalism and empiricism. More about it later. The problem is that Kant's theory comes at the high price of undermining our oldest assumptions on how human ideas about the world match up to the external world itself. For, according to Kant, all knowledge of the external world is a human and therefore mediated knowledge; we can never have knowledge of the world itself, knowledge separate from its knower. Which, of course, is unsatisfying – since that's what we usually mean when we say that our knowledge is "objective," namely, that it corresponds *perfectly* with the external world.

Kant's Critical Philosophy was a significant moment. Among much else, it made possible the two main post-Enlightenment traditions within Western philosophy, loosely designated as "Anglo-American" (or "analytic") and "European" (or "Continental"). The first essentially recoils from Kant's conclusions about the human horizon of knowledge and continues onward with faith in science and a commitment to empiricism, while the second accepts the restricted horizon of human mediation as the proper foundation for understanding the world. The upshot is the opposing traditions of logical positivism and phenomenology, the two very different offspring of Kantianism.

The Democracy of Suffering presumes this well-trodden path, hastily sketched, and asks a handful of questions about our world today: How, indeed, do human *subjects* know the *object* world – including, most pointedly, a world of human-created global warming? Beyond that, what kind of human subject is possible after Kant and the age of Enlightenment? After the horrors of the twentieth century, after Auschwitz? And, getting closer to my target, after the experience of our "postmodern condition"? In short, who are we today in the age of catastrophic climate change, the time of the "Anthropocene condition"?

My approach has been to clear the theoretical cobwebs to come to a better understanding of what is at stake with our present and the future to come. To this end I begin by reviewing in breezy but (I trust) theoretically felicitous terms how past philosophy relates, or fails to relate, to the material, cultural, social, economic, and philosophical realities of our present time and what it might mean to our possible tomorrows. Above all I want to know how we got stuck with this mess, what the Anthropocene condition is doing to our sense of identity, consciousness, and shared humanity, and how it will shape the future of human society.

It's probably clear by now that my starting point is different from what you will find elsewhere in the burgeoning literature on the Anthropocene. But before we can save the world, I think it's important that we understand how we frame and know the world. I have therefore posed questions of our present condition that are quintessentially philosophical. Here's a few more. What does the condition of living in the Anthropocene have to do with the meaning of life? And what does it tell us about the existence of the external world around us? What about nonhumans? Can the experiences of philosophy, especially in the European tradition, give us a handle, language, or gravitas useful enough to help us navigate this new experience?

What follows is exploratory diagnosis and analysis. I leave to smarter and better-equipped people the task of ameliorating the actual conditions of a hotter world beset by crisis after crisis – environmental, economic, social, political, philosophical, human, *everything.* On this score I think we already know, in part because we already feel, that the world of the near future will *not* look like the world of the past. Hume was obviously right to ponder the "problem of induction": future events in the empirical world are in no way obliged to repeat the patterns of the past (see Romm 2018). And they certainly don't, as that dismal science, economics, learned yet again with the market crash of 2008. Prediction, and therefore the fate of the human and natural sciences, has never been more perilous in a world punctuated by "black swan" events – events so rare as to dramatically change what had been, until that moment, the received common sense of a given time. Hence the perfectly modulated term "global weirding."

This brings us to the first major feature of the Anthropocene condition: measured against the relative stability of the last epoch of life on Earth, the Anthropocene is a time of chaos and unpredictability, in which the frequency of (random, unpredictable) black swan events challenge the very meaning of what passes for normal. As the World Bank Group puts it, "climate change brings *deep uncertainty* rather than *known risks*" (2016, 7). It's a time, therefore, more suited to *interpretation* than *prediction.* For uncertainty about the basic conditions of existence means the future has opened up in a way that renders the sciences – based as they are on the ideal of predicting the future – increasingly impotent. In this context humane scientists need to do what they do best: step up and engage, historicize, analyze, interpret, and imagine what comes next.

In this respect I don't think we need more research about the reality of global warming – although good research is obviously useful. Every new scientific report confirms what we already know, only worse. What we do need is higher-order analyses, or meta-analyses, of what the Anthropocene

GLOBAL WARMING

"Global Warming Microclimatic Trend" (2015 diptych), Andy Singer

is already doing to human beings and to the planet. And then we need a plan of action. Why favour interpretation and analysis over prediction and data collection? Because, in my view, we already live in the Anthropocene, the time *after the Holocene epoch*. Because impactful climate change is very obviously the result of human-generated ("anthropocenic") carbon emissions in the world. And because we need new ways of thinking, new philosophies, appropriate to these two facts. In short, the future has in some ways already arrived.

So no more debating science with the (mercifully) dwindling number of critics of global warming. No more stalling as the world turns, the conditions for life on the planet worsen, and people, animals, flora, and fauna suffer and die. No more bullshit. The matter is settled and everyone knows it.

Well, nearly everyone. Those who don't know it by now either don't care or won't care. I propose we stop caring so much about their carelessness, motivated or otherwise, and get to work thinking about our predicament – analyzing the conditions that are preparing the ground for our collective future.

SNAPSHOT 2015

Historians may look to 2015 as the year when shit really started hitting the fan. Some snapshots: In just the past few months, record-setting heat waves in Pakistan and India each killed more than 1,000 people. In Washington state's Olympic National Park, the rainforest caught fire for the first time in living memory. London reached 98 degrees Fahrenheit during the hottest July day ever recorded in the UK; The Guardian briefly had to pause its live blog of the heat wave because its computer servers overheated. In California, suffering from its worst drought in a millennium, a 50-acre brush fire swelled seventyfold in a matter of hours, jumping across the I-15 freeway during rush-hour traffic. Then, a few days later, the region was pounded by intense, virtually unheard-of summer rains. Puerto Rico is under its strictest water rationing in history as a monster El Niño forms in the tropical Pacific Ocean, shifting weather patterns worldwide. (Eric Holthaus 2015)

An uncontrolled experiment is underway on a global scale, and human beings are both its cause and its future winners and losers. Note that this is not exactly a prediction or forecast, just a statement of fact. What the *realization* of this fact is doing, and will continue to do, to our humanity is an interesting conundrum. Will it make for better or worse people? Better or worse societies? And better or worse in what ways? Will the recognition of the Anthropocene condition inspire despair and economic exploitation or change and revolution? Fascism or socialism? What lies ahead in the very near future for humankind? These are of course the biggest, most far-reaching questions possible, and it is my charge, probably foolishly, to pose and then sketch answers to them.

Philosophy has always asked the Big Questions, even when the answers are not obvious. But today I am proceeding as much as whistle blower as philosopher, someone who has been slowly radicalized by the reality of our present situation – someone, by temperament and training, more comfortable reading, writing, and teaching about the intellectual puzzles of history. Yet desperate times call for a break in old patterns, certainly a break from business-as-usual, and perhaps some risk-taking across the disciplinary silos that divide us. In this mode I am posing what I consider to be the most pressing, most audacious, and most important question of our time and perhaps of all time: *What is the Anthropocene condition?* Moreover, what does it have to do with human existence? It's the grandest kind of question about life and survival in a time of death and extinction, about existence in the face of perceived nonexistence.

Of course the enormity of the question is part of the problem. But it's also, I submit, the very measure of its ability to move us, to awaken us, to demand our attention, and, by extension, to change us and change the natural world in which we live.

Or else.

Or else – what? It's highly tempting to say, with a world-weary shrug, *fuck it*. In fact, fatalism of this sort has been the dominant response in the world so far, especially in the West. And it's worrisome. What will happen if we continue along the familiar path of environmental, social, and economic destructiveness? There's no mystery here. Within a few decades existence everywhere will grow harder and meaner. In a rehearsal of Thomas Hobbes's famous dictum about the "state of nature," only now applied to the other end of history, life for many in the Anthropocene will become solitary, poor, nasty, brutish, and short. The world will continue to be divided into lucky and unlucky, wealthy and poor – only more so.

The lucky will enjoy the necessities of life, like water, food, and security. The unlucky will not. And despite what most commentators say, this condition of inequality will accelerate most dramatically at home, within Western society, which after all has the most to lose: that is to say, the farthest to fall. Since poverty, misery, and death are already the lot of much of the world, the loudest complaints will come from the newly disadvantaged citizens of the West – although of course the world's poorest people, as always, will die in tremendous numbers.

If so, then we have a second major feature of the Anthropocene condition: inequality will continue to grow, and grow massively, not between the West and the rest of the world, but most of all within Western society itself. As features go, it should be the least surprising. Rising inequality is already an open secret as the political and economic elites are busy cultivating a "Third World" at home (see Temin 2017), slowly undoing the good luck of being born, for example, in the United States or France.

THE ELITES

The barbarians aren't at the gates. They're already here in the boardrooms; they've been here all along.
(Jacob Bacharach 2018)

This is Canadian journalist Naomi Klein's perceptive argument about the disastrous spread of the Green and Red Zone logic of war, as experienced on the ground in Baghdad during the Iraq War, into everyday American

Refugees, courtesy of Pixabay

Greed, courtesy of Pixabay

society. She points to Sandy Springs, a wealthy enclave of Atlanta, Georgia, that in 2005 separated from the surrounding community to form its own (let's just say) corporation – thereby withholding (hoarding) its taxes from the surrounding community. To effect this change they hired a private company, CH2M, once contracted to manage affairs in the Iraq War, thus closing the circle between the conditions of war and peace. Other cities, like Milton, quickly followed suit.

The greed and small-minded meanness of this trend are shocking; more so because their effects are totally predictable. "The partitioning of the country," Klein observes, "would create a failed state on the one hand and a hyperserviced one on the other" (2007, 508). And that's just it. It's not just one community being destroyed by rampant neoliberalism, American free enterprise run amuck. It's happening everywhere. As Peter Frase puts it, "Gated communities, private islands, ghettos, terrorism paranoia, biological quarantines – these amount to an inverted global gulag, where the rich live in tiny islands of wealth strewn around an ocean of misery" (2016, 129). Then he points to plans (reported in *The Guardian*) to build a private sustainable city for 250,000, Eko Atlantic, off the coast of Lagos, Nigeria. Such "climate apartheid" is the inevitable result of a capitalism unmoved by the plight of anyone but the very wealthy.

Here's a modest prediction: If such activity doesn't stop very soon, the United States will be unrecognizable in twenty-five years. Already an oligarchy in our lifetimes, this once-great democracy – having entered what the Pentagon (in *At Our Own Peril*) openly calls a "new period of post-US primacy" (2017, 3–5) – will degenerate into a full-blown military state characterized by Green and Red Zones, safe zones and "sacrifice zones." Its biggest business will be "carceral capitalism" (Wang 2019), to wit, the jailing, killing, and confining of its own, by then dispossessed and disposable, citizens. The cascade of events is as predictable as it is horrific: unwarranted harassment by police leads to a ticket, unwarranted searches, jail cell, loss of a job, impoverishment, and so on. The arbitrary exercise of power is not only unjust but absurd – or Kafkaesque. But that's essentially how life already is for some people today. It's also a cautionary tale of how things could become for more of us in the future. If so, the biggest business will be war against all enemies – beginning with the state's own visible minorities, poor, homeless, and criminalized citizens, and extending to those homegrown "terrorists" called activists, protestors, artists, intellectuals, and journalists. Such is the logical endgame of a system that promotes the militarization of local police forces and, along with it, the rise of SWAT raids from hundreds *per year* in the 1970s to over one hundred *per day* in 2015 (Frase 2016, 137).

WHO'S A TERRORIST?

Hundreds defied thunderstorms in Ottawa on Saturday, May 30 [2015], and peacefully marched against Bill C-51, Prime Minister Stephen Harper's "secret police" legislation. And then there was a revealing exchange between a group of protesters and a Royal Canadian Mounted Police (RCMP) officer.

The moment, which was caught on camera, suggests that some in the rank-and-file of our national police force may be struggling to distinguish between legitimate dissent and terrorism.

Asked about his opinion on Bill C-51, the Anti-terrorism Act, 2015, the RCMP officer responded with: "Whenever you're attacking the Canadian economy you could be branded a terrorist." (Madondo 2015)

Truly, it's as though our most recent experiences of everyday reality have been swapped out for an apocalyptic science fiction movie, only running in real time or, if you're lucky, in slow motion. The speed of change depends on where you live, Malè in the Republic of Maldives, New Orleans on the eastern coast of the United States, Fort McMurray in Northern Alberta, the city of Athens in Greece, or Tokyo, Japan. The real question is whether or not the masses will accept their role in this zombie apocalypse of our own making or will rise up and demand change. And whether or not the political leaders will continue to allow the worst things to happen to the very people they are elected to serve.

"Just for the sake of security and to keep the Convention proceedings orderly, lets restrict press and protester access to 1,200 feet."

"Conventional Whizdom" (2016), Dwayne Booth

Governments and corporations are not just failing us, they are the driving forces that are taking us to the brink, wilfully ignoring the consequences and thereby committing what can only be called an intergenerational crime. (David Suzuki 2013)

So yes, it all sounds grossly hyperbolic, like a thought experiment gone wrong. Yet it isn't merely alarmist (and highly entertaining) fiction. The truth is that dire scenarios of governments cracking down on their own people barely count as prediction. Once again, it's just a *description* of a future that already exists in shocking abundance today. Jailing, killing, and confining is now as American as apple pie – and much more profitable. Just ask Dick Cheney and Blackwater. Or, better, ask the taxpayers of California, who were greeted with the following headline from the *Los Angeles Times* on 4 June 2017: "At $75,560, housing a prisoner in California now costs more than a year at Harvard."

"GLOBAL RESEARCH" STUDY IN 2008

The figures show that the United States has locked up more people than any other country: a half million more than China, which has a population five times greater than the US. Statistics reveal that the United States holds 25% of the world's prison population, but only 5% of the world's people. From less than 300,000 inmates in 1972, the jail population grew to 2 million by the year 2000. In 1990 it was one million. Ten years ago there were only five private prisons in the country, with a population of 2,000 inmates; now, there are 100, with 62,000 inmates. It is expected that by the coming decade, the number will hit 360,000, according to reports. (Cited in Peláez 2008)

If this description is true, or even half-true, then immediate change isn't really a choice. Or it shouldn't be. It's a moral and intellectual necessity. The alternative is to stay the course, chillaxing at home and surfing the net while venturing out now and again to restock supplies at the nearest temperature-controlled shopping centre. Or, if you're feeling lazy, have a package delivered by drone. Shop till you drop! Shop so terrorists don't win! Shop for democracy! Shop for freedom! Shopping as subversion! Shopping as warfare by other means! Madness has never seemed so inviting and, indeed, so enjoyable. But surely these are the scariest illusions of all.

SHOPPING FOR FREEDOM

"Get *down* to Disney World in Florida," he [President Bush] urged just over two weeks after 9/11. "Take your families and enjoy life, the way we want it to be enjoyed." (In Becavich 2008)

•••• ••••

This little book is divided into three sections, each associated with rough start-and-finish dates, corresponding to a discussion of the *past* in philosophy; the *present* in the social, economic, and natural environment; and the *future* as found in clues all around us and through the exercise of imagination. The three moments follow, one after the next, but readers are invited to start wherever their interests lie most – jumping straight into the future, perhaps, as a way of confronting the past or beginning with the present as made possible by the birth of the Anthropocene. To assist readers I summarize the territory just covered at the end of sections 1 and 2.

In the first section I turn to Kant and Enlightenment as a useful prelude to pondering the condition of our most recent postmodernity, so that we can better understand our present and its possible futures. Here and elsewhere I also consult the mature Plato of the *Republic* – really the first thought experiment, or science fiction scenario, to imagine the ideal society. It's probably the most intellectually challenging of the three sections.

In the second section I turn to the rise of capitalism, globalization, and neo-liberalism in the "postmodern condition" and then consider the moment we self-consciously stepped out of the relative comforts of the Holocene epoch. These are the immediate birth pangs of the Anthropocene condition. In the third section I turn to the future, in part to understand the limitations of the past and present but more specifically to explore how our philosophies and environmental thinking could cash out as significant social change. What follows, in short, is a reflection on three stages in recent human history:

1. The human as subject of reason and Enlightenment (the past);
2. The world as object of capitalism (the present); and, finally,
3. The postcapitalist world being generated out of the dialectic or encounter between both, that is, between human beings (subjects) and the Earth (object) upon which we live (our collective future).

To add one diabolic element, I have rendered the "present" as the period from 1968 to 2008 (I'll return to this later) and the "future" as the period from 2008 to 2100. True, the Anthropocene Working Group of the International Commission on Stratigraphy has yet to decide, formally, if we have indeed entered the Anthropocene. But I humbly consider the matter settled – not scientifically, of course, but socially, culturally, and intellectually. In this way I am telegraphing my view that we are already living in a future defined by the Anthropocene condition – which is precisely why aspects of this future are apparent to us today when they were not just fifteen years ago. In short, I think we are living in epochal times.

JAMES HANSEN ON THE FUTURE

I think on the shorter term, the planet becomes much less habitable – low latitudes become less habitable, and if we lose coastal cities everything starts going backwards. The progress we've had over the last centuries, more people having a higher standard of

living – that's going to go in the other direction. So we really have to stabilize it at a level that allows ice sheets to remain on Greenland and Antarctica with sizes comparable to what they have now. And that requires that the warming be not more than a degree or so.
(In Wallace-Wells 2017b)

In these three sections I wrestle with a cast of brilliant, influential, and sometimes controversial thinkers and commentators working within the Western tradition. These include Plato, Aristotle, Hegel, Marx, Nietzsche, Freud, Heidegger, Adorno, Herbert Marcuse, Michel Foucault, Hannah Arendt, Jean-François Lyotard, Richard Rorty, Fredric Jameson, Mark Fisher, Naomi Klein, Stephen Gardiner, Denis Cosgrove, Francis Fukuyama, James Lovelock, Donna Haraway, Bill McKibben, Elizabeth Kolbert, Cary Wolfe, Jedediah Purdy, Kelly Oliver, Roy Scranton, Dipesh Chakrabarty, Nick Srnicek, Franco Berardi, Alex Williams, Imre Szeman, Wolfgang Streeck, Jacques Derrida, Walter Ong, and many more. My claim is that thinking about critique, Being, consciousness, identity, subjectivity, and labour and thinking about the key milestones of Western thought, such as Enlightenment, humanism, counter-Enlightenment, postmodernity, antihumanism, and posthumanism, while paying attention to the contest between capitalism and its critics and the subsequent emergence of post-capitalism – myriad features of the Western experience – enables concrete insights into the dawning of the Anthropocene condition, an entirely new moment in the history of human consciousness, and its role, whether we like it or not, in generating the world of tomorrow.

At stake is the birth of a new kind of human subject in a new kind of environment, one hitherto unknown to human history. At stake, in short, is the future of humanity itself.

The Democracy of Suffering is a short, accessible introduction to the climate change debate and its literature, and doesn't pretend to be comprehensive or exhaustive. Interested readers should raid the bibliography for authors that strike them, and dig in where I have not.

THE DEMOCRACY
OF SUFFERING

PART ONE

THE PAST
(CA 1784–1968),
SUBJECTS OF
REASON

1 Norms

Philosophy is always determined by transient social and political norms, either as a reflection of power or as a beacon for change – or as a hybrid mixture of both.

Philosophy is a reflection of power when, as in Aristotle's politics or Hegel's history of human consciousness, a philosophy is used to rationalize a *current state* and its prime representatives. If such thinking moves toward an end, that is only because it begins, conceptually, at the end: an end that is already known, namely the "present." The purported beginning of analysis is actually, therefore, a secondary (deductive) feature of retrospection, of having worked one's way from the known present all the way back to the unknown beginning. Which is to say the "discovered" origin is actually the aim or purpose of such thinking, the time-before that rationalizes norms associated with a given present (the "end").

Philosophy is a beacon for change when, as in Plato's theory of the ideal state or Marx's theory of communism, a philosophy is used to rationalize a *future state* and its prime representatives. If such thinking moves toward an end, that too is because it begins, conceptually, at the end – an end that is unknown because unknowable, namely the "future." The purported end point of analysis is actually, therefore, an inductive feature of anticipation, of having worked one's way from the known beginning, the present, all the way up to the unknown end. Which is to say the end as future is actually the purpose of such thinking, the time-ahead that rationalizes the unrealized norms of our present world. These unrealized norms are usually called *ideals*.

This second, anticipatory or aspirational kind of philosophy is obviously the more precarious, since it reaches beyond what is given to the present and, like an invocation, imagines or gives what is yet-to-come, what is barely given over to thought. It is the primary reason why Plato's thinking about politics, so incredibly free of contemporary Greek norms, is in many respects laudable while Aristotle's is not. For Aristotle's retrospective philosophy is always correct, as it were, by definition – a self-fulfilling quest for an origin that merely rationalizes the given present. And so, for example, Aristotle's rationalization of the fully realized privilege of propertied male citizens risks nothing to gain everything, while Plato's anticipatory philosophy based on intellectual and physical merit risks everything to gain everything, although in practice usually gains nothing. That's why Aristotle seems so sensible even when he's clearly wrong, and Plato seems nonsensical even when he's clearly right.

The truth is that we hold Plato and *prospective philosophy* to an entirely different standard than Aristotle and *retrospective philosophy*, since no one expects ideals to be realized – least of all Plato. That's because the heuristic power of prospective philosophy issues from those very unrealizable models and measures of an ideal future; issues, that is, from an absolutist and infallible perspective of moralism that one wields in the present like a whacking stick. Not so Aristotle, whose thinking was fatally dated the moment his retrospective rationalizations failed to align with the ever-changing political times. Aristotle's hardheaded realism was rendered merely arbitrary with the passage of time, while Plato's aspirational fantasies have been renewed with every tomorrow.

If I pause over these niceties it's because humanity has arrived at a propitious moment in history. First of all, it's almost certainly a bad time for life on the planet. Second, there is a very modest, mildly perverse silver lining: it's an especially good time for philosophy, the "love of wisdom." Why is that? Because the sorry condition of the planet is impelling philosophy to reconnect with its present as understood, not as simply determined by its past but as inspired by its possible futures. Here's my basic claim: dire times call for a more laudable, aspirational, moralistic philosophy based on conjuring, imagining, and thinking the unknown – risking everything to gain everything, even while courting nothing – according to still unrealized norms. Only such a philosophy can pick up the whacking stick of the future and put it to work today.

As for retrospective philosophy, which, in the extreme, functions as a proxy for power, it's unequal to the enormity of the task of thinking today's present. It is too busy making the past into an excuse for our present, a present that has arrived at an impasse. Less charitably, this kind of philosophy has been rendered metaphysically and morally bankrupt, dead.

2 Kantian Subjects

The form of the question, "What is the Anthropocene condition?," recalls a question originally put to Kant and his contemporaries in the eighteenth century: "Was ist Aufklärung?" "What is Enlightenment?" Let's begin by simply echoing Kant's circumspect answer to the question of Enlightenment.

It is, he claims in the opening paragraph, humankind's emergence from self-incurred immaturity; such immaturity is self-incurred if it inclines people to submit their reason to the authority of another. The motto for Enlightenment is therefore, Kant concludes, *Saper aude* – Dare to know – a phrase, lifted from Horace, advanced as a motto for the Society of the Friends of Truth decades before in 1736.

While Kant doesn't ignore the idea that Enlightenment is the "Age of Reason," he doesn't belabour it either. That is, he doesn't answer the question of Enlightenment by reciting the dominant characteristics of an *epoch*: for example, a belief in social and intellectual progress, the perfectibility of human nature, the universality of reason, and the role of science therein; a new faith in and optimism about the future; and the rise of democracy, freedom, and human rights. Instead Kant refers to the dominant characteristics of an enlightened *subject* – a subject defined as mature, reasonable, responsible. In other words, Kant doesn't begin with the historicity of "the Enlightenment," with easy periodization, but provides a précis on subjectivity. The Enlightenment promises a future, close at hand, when the mass subject will share in the heady power of science and reason, and also in its responsibilities as steward. The mechanism of change, the ladder, is the education of regular people, an idea inspired by Jean-Jacques Rousseau, and the reward is the maturation of a society yet to come. In this way Kant anticipates the eventual triumph of culture through the massification of mature, rational, duty-bound individuals.

Kant, we know, predicted cosmopolitanism and hoped for perpetual peace. But what we actually got from his future of universal mass culture was the secular, profane late twentieth century. We got Americanization and globalization. And now, at the dawning of the twenty-first century, we've got something else again: cynicism, fatalism, retribalization, populism, austerity, and the groundwork for future wars.

3 Subject of Change

The poet Heinrich Heine famously credited Kant with being the "philosopher of the French Revolution." But that's a stretch. It's certainly true that Kant sides with revolution over traditionalism, with Thomas Paine and "the rights of man" over Edmund Burke and the rights of tradition. As Kant argues in "Answering the Question: What Is Enlightenment?," one generation cannot shackle another generation to imperfect and irrational policies from the past. Progress demands that social and political norms, inherited from our dead ancestors, be renegotiated with every new generation.

Even so, Kant was hardly the philosopher of violent overthrow, of violent revolution. He never advocated beheadings. His call was for careful, reasoned, dutiful criticism of the existing social and political conditions of life. And so he pointedly echoed the dictum of Frederick II, his Prussian king: "Argue as much as you like, but obey." Or as he bluntly says elsewhere, human beings everywhere "require a master." Kant was at best, therefore, a reluctant revolutionary – the philosopher-as-midwife to the future. The far less genteel role of philosopher-as-undertaker was left to others, left to the future, left most conspicuously to Karl Marx, Friedrich Nietzsche, and Sigmund Freud.

Just as Kant refused to reduce the Enlightenment to a proper noun, philosopher-midwives must take care not to make a thing, entity, or fetish out of our own momentous time, "the Anthropocene," the era of human-generated climate change. That said, the truth is that the Anthropocene condition punctuates and, by comparison, dwarfs the event called Enlightenment, transforming the rational, conceptual, and abstract into something else altogether: the concrete, geological, and earthen. More about that later. Enough for us now to simply register that imagination falters before this transformation. For this reason I am beginning at the beginning, in full retrospective mode, by insisting upon a properly Kantian question: what does our present time have to do with the future of human *subjects*? What has any of it to do with what philosophers call "ontology," with being? More simply put, what does it have to do with you and me?

It is perhaps not too cheeky to say that everything revolves around this question, not just the Earth upon which we stand but the limits of human knowledge that Kant defines and then stumbles against in his "critical philosophy." In this respect it's not for nothing that Kant describes his own philosophy as a "Copernican revolution" in philosophy, a dramatic realignment of perspective. For Kant was the thinker to realize that the objects of knowledge are necessarily framed by our limited human reckoning. As such, Kant's revolution was overwhelmingly ironic. For within a period of roughly 200 years Westerners were obliged to grapple with a paradigm shift, the scientific revolution, that displaced human beings from the centre of our cosmology, only to be told by Kant that knowledge of the physical world does indeed revolve around human beings, around knowing subjects. One could therefore speak of Kant's Copernican revolution as a counter-revolution: the world is known only through our conceptions of it or, if you prefer, the object world rotates around the conceptual limitations of knowing (human) subjects. As the faithful Arthur Schopenhauer asserts as the starting point of *The World and Will and Representation*, and in loaded quotation marks: "Die Welt ist meine Vorstellung," the world is my idea, my representation of it (1819, 3).

Since Kant, it has been ever thus – at least in the phenomenological tradition he inspired. And so a third feature of the Anthropocene condition – it is a kind of *gross literalization* of Kant's insight into human reason and seems, as such, like a suitable foundation for realism. *The object world that we know is in fact a human world.* Human beings *make* the world. Or again, the external world is a *direct or unmediated* reflection of human reason, imagination, ceaseless activity. It follows that the Anthropocene is all about human subjects and the world they stamp as their own. In short, the object world *is* the subject (or human) world.

4 The Anthropocene Condition or "Anthropocenity"

Let's float and then define, tentatively, a new word that encapsulates the spirit of our time, one that functions analogously to the word "postmodernity": *anthropocenity.*

Anthropocenity: an abstract noun from the Greek ἄνθρωπος, ánthrōpos, meaning "human," while "cene" means "new," specifically as the term refers to a new period within the Cenozoic era. The "ity" suffix telegraphs the not inconsequential *condition* of living in this new human era. And sure enough, at least one other author refers to it in these exact terms: "We can call the sum of these changes, the vast and irreversible human impact on the planet, the Anthropocene Condition" (Purdy 2015, 4).

Really, the word says it all. As a concept, anthropocenity stamps being and time as *human being* and *human time.* For humanity has stamped the external world itself with a force, literally and figuratively, that is so impactful that it can with justice be called "geological"; our radiation and plastics, for instance, are literally registered around the world on the geological level. Thus we could say that, under the Anthropocene condition, time itself has collapsed, becoming an extension of human imagination – becoming

2.0 and 2.1 "Flames and Cranes 1 & 2" (2016), Maria Whiteman.

human. Or, if you prefer, we could say that human being now coincides with time. The time of "anthropocenity" is the time of human-change. Such is the collapse, according to the historian of postcolonialism Dipesh Chakrabarty, of "the age-old distinction between natural history and human history" (2009, 201). As a consequence, human beings have for the first time, he argues, become "geological agents" (207). Chakrabarty says it produces the affect of "falling" – for example, of "falling into deep or big history" (2015, 181), which is to say, of unsettling one's present time in the vastness of "species history" or "species thinking" (2009, 212–13) – even as it registers a "certain shock of recognition," most especially the "recognition of the otherness of the planet."

This upsetting of human existence, this identity crisis, is all the more remarkable in an era so interested in nonhuman rights, from the ontological status of animals all the way, with speculative realists, to the ontological status of objects, things, and even concepts in the world. Yet if, as some contend, we have finally stepped "beyond humanism" – the measuring of all against the exigencies of humanity – that shouldn't be mistaken for a step *beyond* the human subject. On the contrary. Never before has the human subject been recognized as a force equal to the planet. Never before has the world been so totally stamped as "human." The step beyond humanism, so-called "posthumanism," is therefore a question of the future. But the essential question of the Earth will always remain a question of the human subject – of you and me. Back to this in sections 2 and 3.

> Humanism (noun): a doctrine, attitude, or way of life
> centered on human interests or values; *especially*:
> a philosophy that usually rejects supernaturalism
> and stresses an individual's dignity and worth
> and capacity for self-realization through reason.
> (Merriam-Webster)

5 The Critical Subject of Power

According to French philosopher Michel Foucault, Enlightenment is the "first epoch to name itself" (2007, 86) and "formulate its own motto" (87). In a sense the discipline of philosophy had, for the first time, approached the territory of journalism, a grappling with the present, *l'actuel*. It makes sense, in turn, that the question of Enlightenment – which, Foucault claims, is "even more remarkable than the responses" (121) – was posed by and for a Berlin newspaper. And so he writes:

> The Prussian newspaper was basically asking: "What just happened to us? What is this event that is nothing other than what we have just said, thought and done – nothing other than ourselves, nothing other than this something that we were and that we still are?" (121)

Much is at stake in these late musings of Foucault about the "power–knowledge nexus" made possible by the event of Enlightenment, not least of which is the future of "Foucauldianism" itself, which is to say, of the approach he devised called "discourse analysis." Foucault seems to align his own constellation of terms – "archaeology," "strategy," and "genealogy" – with Kant's terms of "critique," "revolution," and "Enlightenment." In this way Kant's terms operate as subtle precursors to, and proxies for, Foucault's terms, revealing a continuity between Enlightenment and the postmodernity of the post–World War period. In this way we might say that Kant's thinking, most especially about criticism and critique, *made Foucault possible*.

[T]he thread that may connect us with the Enlightenment is not faithfulness to doctrinal elements, but rather the permanent reactivation of an attitude – that is, of a philosophical ethos that could be described as a permanent critique of our historical era. (Michel Foucault 2007, 109)

In lectures and interviews delivered between 1978 and his death in 1984, Foucault's analyses of critique, revolution, and Enlightenment exhibit an awareness that discourse analysis, not unlike the "culture critique" associated with the Frankfurt School, is vulnerable to the charge of undue pessimism. Many felt, and sometimes still feel, that the overwhelming influence of disciplinary power in Foucault's account of human and social reality makes resistance and agency – human freedom – impossible.

By way of Kant's little essay on Enlightenment, the late Foucault wants to show how resistance and change is still possible in the face of "governmentality," as instituted, for example, in hospital and penal expressions of power. Foucault's basic answer is that an attitude or ethos of *critique* is built into Enlightenment rationality from the outset, that Enlightenment includes the critical idea that we should be governed differently and can be free in new ways. The criticism of power is, in short, already a key feature of the power–knowledge nexus. We are free to play within, subvert, or even redefine the boundaries of power–knowledge. As a consequence, a weak form of agency is possible, even inevitable, although we must abandon, Foucault cautions, the "dream of revolutionary change."

This is the late Foucault of "self-fashioning," of *áskēsis* – the unleashed Foucault who, discovering an exemplary glimmer of creative dissidence in the gay communities of San Francisco, adjusted his thinking about power accordingly. So while Baudelaire held up the *urban dandy*, an ascetic self-inventor, as the embodiment of modernity, Foucault implicitly and provocatively held up the *queer* as the privileged subject under the conditions of postmodernity. Or, at least, this was the future that was to quickly subsume much of Foucault's legacy until the present time.

For Foucault the human subject is a wild card capable of donning masks or identities as our ever-shifting contexts dictate, a claim that became highly influential among postmodernists. Nietzsche's refuge, similarly, was the artist, a subject perfectly at home in a world of artifice: one who makes an art of her own life. For both Foucault and Nietzsche, the marginalized subjects of power–knowledge are the inevitable critics of power–knowledge. For both thinkers, outsider philosophy is brought to bear against insider philosophy.

6 *"The Present as Philosophical Event"*

Foucault (2007, 85) argues that Kant's answer to the question "What is Enlightenment?" is a privileged one, that for the first time a thinker dared to think the question of the present. "The question which, I believe," writes Foucault, "for the first time appears in this text by Kant is the question of today, the question about the present, about what is our actuality: what is happening today?" (2007, 83–4). For Foucault, Kant's call for the education and maturation of the masses is a call for criticism, and criticism as such is connected to Kant's larger project of "critique" – and therefore to the project of critical philosophy. And actually Kant is clear about this connection. In his preface to the first edition of the *Critique of Pure Reason*, Kant writes that "Our age is, in especial degree the age of criticism, and to criticism everything must submit" (in Foucault 2007, 37, n3).

In characteristic fashion, the age of criticism is, for Foucault, tantamount to the birth of a new attitude, if not a new form of subjective being-in-the-world. "It is not an issue of analyzing the truth," Foucault contends, but is instead an issue "of what we could call an ontology of ourselves, an ontology of the present" (95). This shift from truth to ontology (the study of being) has a lot to tell us about our own attitudes toward power, change, and revolution "today," even as it points toward the central claim of Foucault's entire project: the "modern age" made possible a new kind of human subject, a subject of power who was simultaneously aware of power, and thus given over to the critique of power and the power of critique. A subject *and* object of power–knowledge.

If, however, there's a question that remains largely unformulated in Foucault, it's the question of the future – a question asked most insistently by Nietzsche and, in turn, by Heidegger. Foucault denies that the future motivates Kant on the question of Enlightenment. Kant, he says, "is not seeking to understand the present on the basis of a totality or of a future achievement. He is looking for a difference: What difference does today introduce with respect to yesterday?" (99). But Foucault is wrong. Kant always presumes the future of his own present, something that becomes very obvious in his "Idea for a Universal History from a Cosmopolitan Point of View." (More about it in a moment.) It is therefore inadequate to say that Kant was "the first philosopher to think his present time," the Enlightenment, as it was reflected in his own philosophy. For even if we grant Foucault this (not entirely convincing) claim about the uniqueness of the journalistic present for Enlightenment philosophy, the Kant that thinks the conditions of his own actuality – the "conditions that made him possible," in Kantian language – this Kant is thinking about conditions as determined by a future. That's largely because Enlightenment, as Foucault himself admits (97), is for Kant still an *open* question. And by "open" it is necessarily "oriented toward the future."

If Foucault reduces the Kantian orientation to his own, it's because Foucault himself is so committed to the present. Foucault admits, for instance, that his critique of madness fits hand-in-glove with the antipsychiatry movement in the English-speaking world during the 1960s (137). In this sense his historical research was motivated by concerns ripped from the pages of newspapers. Both Foucault and Kant are investigating the parameters of what it means to be a "public intellectual." But Foucault's philosophy, like Aristotle's philosophy, is really a philosophy of *retrospection*. Obviously his aims are very different from Aristotle's. Aristotle looks back to demonstrate that his present is correct, natural, inevitable, final. Foucault looks back to demonstrate that his present is, perhaps not incorrect, but artificial, constructed, and cultural – not inevitable or final but radically contingent. Aristotle reifies power, unity, sameness. Foucault reifies powerlessness, disunion, difference. In other words, one could say that Foucault's retrospective philosophy *inverts* Aristotle's rationalization of present-day norms, making an end, not of the citizen and philosopher, but of the disowned and mad of everyday contemporary life.

That Foucault thereby makes an antihero of the outsider is clearly very important from a personal and ethical perspective. But his retrospections are committed to the same operation we find at work in Aristotle: the past is found (or made) to justify the present. Both Aristotle and Foucault *use* history, and use it selectively, to understand the present and to that extent do not escape from history.

This is not a criticism – or only a criticism. It is merely admitting that the horizon within which Foucault worked is history as determined by the present, his present. It is indeed a "history of the present." Foucault's thought was never a horizon determined *by the future*. No doubt the disappointing failures of 1968, of the revolutionary future, took the wind out of the sails of nearly every French intellectual of his era. As a consequence many thinkers, including Foucault, retreated into the past for answers.

Once again, this isn't Kant's approach. Kant's thought is very much determined by an event unfolding in his own time but the destiny of which remained to-be-seen, i.e. remained for the future. Kant's answer to the question "What Is Enlightenment?" is oriented toward this future, toward this destiny; it is not an historical answer to a present identity or politics. Indeed, Kant's answer – a chronicling appropriate to its venue, the newspaper *Berlinische Monatschrift* – was anticipatory in its tone and style. Arguably that anticipatory tone is still instructive today as we face the uncertain future in the Anthropocene. And that means that Kant is still instructive.

7 "The Profound Destination of Western Reason" (Foucault 2007, 52)

Kant's "What Is Enlightenment?" was published in 1784, the year between the publication of the first and second editions of the *Critique of Pure Reason*. This is the time of the later, fully mature Kant. In 1784 Kant also published the "Idea for a Universal History from a Cosmopolitan Point of View," a second intervention that year in the public sphere. As always Kant practised a duty that he preached as a philosopher: to be consciously engaged in the grand project of improving human society. He was a mature, enlightened subject – philosopher as public intellectual.

"Idea for a Universal History" is a model of clarity and, for many readers, a prophetic call for the kind of "teleological theory of nature" – a prospective narrative of history – later taken up by Karl Marx. Kant's argument, reduced to nine theses, is boldly sweeping about the meaning and destiny of humanity. To this end he rejects the possibility, monstrous in its implication, that Nature (a master word warranting caps) has no plan whatsoever for the human species. For if Nature has no purpose for our collective human history, if Enlightenment follows no general laws, then according to Kant humankind would be following "an aimless course of nature, and blind chance takes the place of the thread of reason." Without a *telos* or purpose, humankind would be a "contemptible plaything" of Nature.

Kant recoils from this possibility and asserts that Nature is always meaningful at the general level, in our case at the level of the species. And so Kant compares humanity with the weather: seemingly arbitrary on a small scale but full of meaningful patterns on a larger scale. So the kind of individual foolishness we see all around us says nothing against the existence of a grander law of Nature, by which he means a law of History. "Since the philosopher," Kant writes, "cannot presuppose any [conscious] individual purpose among men in their great drama, there is no other expedient for him except to try to see if he can discover a natural purpose in this idiotic course of things human. In keeping with this purpose, it might be possible to have a history with a definite natural plan for creatures who have no plan of their own." And so Kant, meteorologist of History, dreams of a future that promises the perfectibility of the human species in the achievement of a moral world order that includes not just peace between individuals, but between nations.

According to Kant both individual and collective instincts of aggression, both "the unsocial sociability of men" and the war of nations, drive the slow progress of the human project. Freud makes the same point, more or less, in his late cultural works, where civilization, life, and love exist as elaborate reaction formations to the death drive, to innate human aggression. But whereas Freud is resigned to a perfectibility achieved only over geological time (as he says at the end of *The Future of an Illusion*), Kant is enthused about a perfectibility that can be achieved over human time. Freud is a pessimist, Kant an optimist. In his letter-cum-essay to Albert Einstein (1932) on behalf of the League of Nations, Freud is therefore sceptical about the prospects for future peace and fellowship among human beings. It all belongs to a very distant (but "not infinitely distant") future. But Kant, in 1784, is highly enthused about the near future in which a "league of nations" (he names it explicitly) will ensure a higher level of civility, and enable the true potential of human beings. Such are the "cosmopolitan conditions" that are, Kant believes, the destiny of Nature's design.

Note once more that Kant's teleology is unlike Aristotle's, precisely because Kant's thinking is anticipatory and prospective. Kant is not trying to rationalize the present on the basis of the past but to actualize an unknown but promised (and promising) future. Kant's belief in Enlightenment, in the education of the masses, is a belief in the future as revealed to us by Nature over the course of historical time. In this sense, Marxists are obviously right to see in Kant a fellow traveller committed to a better tomorrow. So Kant, the thinker of a better future, really is in some ways a revolutionary.

8 Mind the Gap

In *The Phenomenology of Spirit*, Hegel recounts the moments in history when self-consciousness becomes certain of itself, that is, becomes recognized and therefore objectified through a "life-and-death struggle" for prestige (1977, 114). The dominant consciousness becomes "master." What follows, however, is Hegel's famous reversal: the master, satisfied in her mastery, eventually becomes dependent upon the loser, the slave; while the slave, at first shaken by the loss of prestige, labours in the world for the master and finally achieves an independent consciousness. The reversal is an irony: the master is enslaved, while the slave is set free. Hegel's grander lesson is that without the slave's labour, without a force of active negation (of reaction), History comes to a standstill; it becomes merely satisfied and, therefore, complacent, dead. The master's self-satisfaction is not yet the actualization of absolute knowledge and not, therefore, the culmination or "end" of History. As Alexandre Kojève puts it, truth has not yet become the "'Science' or the 'System' of the Wise Man" (1969, 173). Or, in Kantian terms, the master's consciousness is not yet a cosmopolitanism appropriate to the unfolding truth of Nature. The master's truth remains a necessary but finally incomplete moment: good until actually achieved, then just a drag on the potential of History.

Marx famously "turns Hegel on his head," that is to say, shifts the field of dialectics away from ideas, philosophy, and consciousness and toward material reality and economics, away from the inessential "superstructure" and toward the essential "base." As Marx says in *The German Ideology*, "Life is not determined by consciousness, but consciousness by life" (1939, 15). The point of History for Marx is not, therefore, the absolute *knowledge* of the wise man but the absolute *freedom* of the people. This humanistic, anthropocentric, or "anthropological" rereading of Hegel, championed by Kojève in the 1930s, was achieved by literalizing the master/slave dialectic, according to which the bourgeoisie is destined to be overcome by the proletariat – and therefore make for a better, more equitable society. And to this end Marx simply *reads* what Hegel says about the value of work: "Through work," Hegel writes, "the bondsman [slave] becomes conscious of what he truly is" (Hegel 1977, 118) or, again, "it is precisely in his work … that he [the slave] acquires a mind of his own," to wit, becomes fully conscious (119). The logic of dialectics, as Kojève rightly surmises, was in other words always "an ontology or Science of Being" (1969, 170). Once again, it's about you and me – human subjects.

But even as the early Marx sharpened his thinking by engaging with the Hegelian dialectic of consciousness, he was also, almost by definition, engaging with Kant's thought about how human beings can know the external world. Let's just review it quickly. Kant finally argues that we can be certain about our knowledge claims about the world, taking refuge in shared "categories of mind" (like causation, negation, and existence) that are given at birth, innate. Yet this partial appeal to rationalism (reason before experience) pretty much ensures that, for Kant, all knowledge remains a human, mediated, and limited knowledge. We can never know the "thing-in-itself" or, more simply, the external world as imagined by naïve empiricists (experience before reason). The world is always represented *by* human beings *for* human beings.

Hegelianism is the attempt to surmount the access problem to *objective* knowledge – not by walking away from Kantianism but by doubling down on Kant's problem with the nascent field of phenomenology. In this way Hegel claims to bridge the phenomenal–noumenal gap (subject–object gap) that Kant, with critical philosophy, leaves as a trade-off for having resolved the differences, the conflict, between rationalism and empiricism, between Cartesian doubt of the external world and Lockean faith that the world of stuff impresses itself directly onto the *tabula rasa* of our minds. The bridge is the History of human consciousness or the "science of experience" – that is, the "phenomenology of spirit," the title of Hegel's great tome.

According to Hegel, Kant fails to appreciate that History itself is the meaningful revelation of the object world to human consciousness; that History itself is the dialectically charged resolution (or "sublation") of all conflicting philosophies into a grand unified vision, a vision nearly achieved in Kant's own critical philosophy. Had Kant recognized (as a phenomenologist) that the historical process that made his own thought possible is *itself* meaningful, he would have known something that was left to Hegel to realize: namely, that human consciousness (or *geist*, spirit or mind) had indeed accomplished its goal and realized a perspective that was beyond all further resolution. History itself had finally been *recognized*; that is, realized as absolute knowledge. The process of coming to know the object world had therefore been achieved, as Hegel put it, because "the real is the rational, and the rational is the real." Reason and consciousness know the world because it *is* the world, i.e., is the perfect representation of it. At last the mind faithfully corresponds to the world. The job of philosophy (to explain how it is that we know the external world) is therefore done.

For Hegel the "end of history" is not, therefore, an achievement of some cosmopolitan future, as Kant believed, but the condition of Hegel's actual lived present. Hegel and the Prussian state had come to embody the truth and meaning of History. As such there was no longer a "veil" at all between mind and external world, no gap between "noumenal" and "phenomenal" realms. In other words, Hegel's retrospective philosophy had neatly rationalized power in and as his own present or, better, in his own presence, his Being.

It's the oldest dream of philosophy: philosopher as the transcendent embodiment of truth. As Nietzsche puts it, this dream has the form of "I, Plato, am the truth."

This doesn't make sense to Marx. For there is obviously still a gap or contradiction at work in History, one known to us as a foundational feature of the material world we live in – as subjects living in the everyday material world of stuff. The resolution of *that* gap is the *raison d'être* of Marxism. So while Hegel thinks he finally *knows* the external world, and has come to embody the destiny of History, Marx argues (in the eleventh thesis on Feuerbach) that the point of philosophy is to *change it* – by an objectively true understanding of the world that will inevitably achieve its destiny in the freedom of all. In short, the destiny of History is not private, individual, and intellectual. It is public, universal, and material.

So yes, there's still a realization or "recognition" at the level of Marxian consciousness. But that consciousness is necessarily collective. As Marx says in a famous remark of 1844, "communism is the riddle of history solved, and it knows itself to be this" (135). In this sense, the communist subject is a "we" – a free objective-subject or, more simply put, a mass subject. It is what Marx calls the "complete return of man to himself as a *social* (i.e., human) being," and again, "it is the *genuine* resolution of the conflict between man and nature and between man and man – the true resolution of the strife between existence and essence, between objectification and self-confirmation, between freedom and necessity, between the individual and the species." Such is *Marx's* phenomenology.

Marx thus shares more with Kant's prospective philosophy than with Hegel's retrospective philosophy. For both Kant and Marx look to a future state to rationalize and fulfill their existing philosophies, whereas Hegel makes of all past states the means to his own perfectly realized present. In a word, Hegel's dialectic remained merely *reactive* and thus incomplete.

To the extent that our present time, the Anthropocene condition, must once again comport itself to the future unmoored, as it were, from the reactive narrative of world History, then perhaps Marx, following Kant, still has something to teach us about our subjectivity in the world. Perhaps the *ontology* of subjects can reveal epistemology. Perhaps "Being" can reveal (a theory of) knowledge. In short, perhaps Marxism can reveal the truth of the capitalist subject.

2.2 "Marketing to Children in Schools" (2006), Andy Singer

9 Being

For Marx, labour implies a theory of being. It's what Marx, in the *Economic and Philosophic Manuscripts of 1844*, calls our "species being." And so the subject fashions the world, for example as a carpenter working on wood, and is in turn refashioned herself. A subject *becomes* what she *does* in and to the object world. Or, again, her essence (inner world) is refashioned through meaningful engagement with the world of stuff (the outer world). As in the case of Hegel's slave, the labouring for the master is transformative, the slave becoming, in its anthropomorphic rendering, a chef, gardener, carpenter, blacksmith, and so on. One's identity is at stake, which is why factory labour under the conditions of capitalism is so alienating – it strips one of (the possibility of creating) an identity forged through meaningful engagement with the object world. One exchanges one's identity for a wage. One is not a carpenter, say, but a wage earner. And being a wage earner is not a concrete realization of one's being in the world. It's an abstraction. One is thus estranged from one's labour, from one's world, and from one's own self.

It's not so much, then, that Marxian human nature is "socially con-structed," as is often claimed, but that human nature is the essential by-product of labouring in the external world of stuff. Our labouring na-ture is itself the bridge between phenomenal and noumenal realms, the missing link that makes the slave an independent consciousness. That is, our labouring nature elevates the human animal, finally in perfect concord with the environment, to the status of *human being*. This is the destiny of History and Marx's answer to philosophy's old question of how a subject can really know the object world: the subject *knows* the world because the subject *is* what she *does to* the world. Knowledge is realized (it is objectified, made real) because it is embodied, lived, acted.

Capitalism foils this process and therefore foils what it means to be human. It is therefore, almost by definition, antihuman and inhuman. Yet Marxism, we know, is not nostalgic for the time before capitalism. It was capitalism that made possible a major leap forward in History. It was cap-italism that made concrete the ancient dream of freedom from meaningless labour, albeit only for those lucky enough to enjoy its benefits. And it was capitalism that made possible, through technology and industrialization and labour, the means for providing sustenance and meaningful existence for all.

The bourgeoisie, during its rule of scarce one hundred years, has created more massive and more colossal productive forces than have all preceding generations together. Subjection of Nature's forces to man, machinery, application of chemistry to industry and agriculture, steam-navigation, railways, electric telegraphs, clearing of whole continents for cultivation, canalisation of rivers, whole populations conjured out of the ground – what earlier century had even a presentiment that such productive forces slumbered in the lap of social labour?
(Marx and Engels 1848)

Conflict between poor and rich, between social (or factory) production and individual (bourgeois) appropriation of wealth, is therefore the final condition of a new era for humankind. It is only *after* capitalism, once the conflicts have been resolved or "sublated," that human beings can achieve an existence equal to the destiny of History, rather than an existence appropriate to animals. This is the essence of Marx's "humanism." As for "communism," it is just the name given to that future when the human animal finally achieves its humanity – its freedom from slavery recast as the realization of a grand human destiny. A destiny perhaps better recast, given the confounding events of the twentieth century, as *communalism* or *collectivism*.

10 On Dialectics

One can readily see just how much of a humanistic thinker Marx really was. So much labouring in the world; so much the transformation of human nature. So much labouring of human nature; so much the transformation of the world. It's not quite the case that "we are the world" and that the "world is us." But that is, for Marx, a close approximation for our collective destiny as working subjects.

Marx's vision of human being and human freedom is a vision of the future. Until freedom is achieved, the gap between subjects and the object world is an itch that can't be scratched, an alienation that robs us of our potential. It's precisely this future – thwarted by the cold war of West and East, which erected a false dualism in place of the living dialectic of history, and thwarted by the unexpected dynamism of capitalism, which sucks all resistance into its orbit – that Marxists have wielded as a club against the conditions of capitalism. A future borne by and through capitalism, but not over or around it, as in the examples of the Soviet Union, China, Cuba, and elsewhere.

One can therefore understand why the critique of capitalism in general is so indebted to Marxism and why the passage to the Anthropocene condition is greased by the wheels of dialectical thought about "History." For only dialectical thought, between idea and action, allows us to ponder, not just the realization of a subject equal to her world but, from the other direction, the becoming-subjective of the object world. And only dialectical thought tells the history of how we got from Enlightenment humanism to "anthropocenity."

11 On the "indefatigable self-destructiveness of enlightenment" (Adorno and Horkheimer 1944, xi)

The alienation of humanity from its essential Being (now rendered in caps) is a cornerstone of Martin Heidegger's thinking about technology, existence, and philosophy. This German philosopher's "One Great Thought" is that Western society has forgotten the truly essential question of Being; that abstract and antihuman metaphysics, ultimately scientism, has eclipsed concrete ontology. This is why Heidegger is convinced that we must vigorously interrogate metaphysical concepts to reveal (or "unconceal") the truth of Being and must bring the history of metaphysics, understood (with Nietzsche) as the history of a life-negating nihilism, to an end. This is the gist of his practice of *Destruktion* – a practice that, in the 1960s, was adopted and transformed into "deconstruction" by French philosopher Jacques Derrida.

Heidegger's early work, *Being and Time*, has been highly influential – even as applied to capitalism and even in the wake of Heidegger's unforgivable support for Nazism. And so, for example, while Herbert Marcuse swapped out the (by 1933) *verboten* project to unconceal Being for Freud's project to psychoanalyze the unconscious, Marcuse never really stopped providing his own *Destruktion* of alienating capitalist culture. Such are the mixed origins of Marcuse's "Freudo-Marxist" critique of technological rationality and instrumental reason, a critique that would become meaningful for the student movements of the 1960s, especially in the United States. For a decade or more, the political revolution promised to be psychosexual – "Marcusean." But it was as much a revolution in consciousness, in Being, as in labour relations and individual psychology. In short, the Jewish Marcuse, appalled by the Nazi politics of his revered teacher, never did shake the old Heidegerian project to interrogate the foundations of "the given."

What one immediately gleans from even this quick sketch is how Enlightenment ideals, in the minds of European philosophers, had faltered and collapsed in the early twentieth century. The twentieth century had revealed a future that was neither reasonable (Kant) nor free (Marx). Enlightenment rationality saw instead rising anti-Semitism and fascism, a World War in 1914–18, the Great Depression that began in 1929, and then another World War from 1938–45. Perhaps most troubling of all, war had morphed into the technologically sophisticated, bureaucratically administered barbarism of the Holocaust. Evil itself, more deadly and efficient than ever, had been rendered "banal" – Hannah Arendt's controversial but perfectly modulated term for genocide in the twentieth century. Mass murder as project management: a "problem" that demands a technocratic "solution."

Theodor Adorno famously wondered how there could be poetry after these horrors, after Auschwitz and the Final Solution, and delivered his dark eulogies for Enlightenment ideals. (A similar sentiment informed commentary in the immediate aftermath of the 9/11 attacks on the World Trade Towers in New York City.) In *The Dialectic of Enlightenment*, published a few years after the Holocaust, Adorno and Max Horkheimer argue that Enlightenment didn't bring universal education to the masses, as Kant imagined, but universal deception; not freedom and truth, as Marx imagined, but domination, totalitarianism, and ideology; not authentic reflection, as Heidegger imagined, but positivistic science; not noble ideals, as Plato imagined, but base efficiency, utility, and instrumental ends; and not a revolutionary future, as every progressive person imagined, but the triumph of the status quo. For the Frankfurt School thinkers like Adorno, the business of dimwitted conformity had become the new normal of

Holocaust, courtesy of Pixabay

twentieth-century Enlightenment. In other words, Enlightenment had turned into its dialectical opposite: "in the successive forms of slaveowner, free entrepreneur, and administrator," complain Adorno and Horkheimer, "the burgher ... is the logical subject of the Enlightenment" (83). A "burgher" is a middle class townie. Updated for our times, she is a suburbanite.

> [T]here is this suspicion that something in rationalization and maybe even in reason itself is responsible for the excesses of power, well, then!: it seems to me that this suspicion was especially well-developed in Germany and let us say to make it short, that it was especially well-developed within what we could call the German Left. (Michel Foucault 2007, 51)

Kant's future realized in the figure of the *soccer mom*.

So much, then, for the Enlightenment subject – and so much for the vaunted ideals of progress and of the future after Kant. As the Italian Marxist Franco "Bifo" Berardi argues, the "utopian imagination" had given way to the "dystopian imagination" (2011, 17). Indeed, the "mythology of the future" was finally over and along with it any belief in the achievement of universal human rights. "The future that my generation was expecting," Berardi observes, "was based on the unspoken confidence that human beings will never again be treated as Jews were treated during their German nightmare. This assumption is proving to be misleading" (19). Cue Cambodia, Rwanda, and Bosnia-Hertzegovina – roughly another three million dead. And cue Syria.

The early twentieth century taught us that enlightened, rational subjects didn't know the external world, hadn't realized and fulfilled the promise of History, but had only succumbed to an ideology that stamped itself on ever-more confused, distracted, and irrational subjects. This was Heidegger's nightmare scenario – a humanity cut off from the truth and ground of Being, measured instead according to the achievements of technical proficiency and utility. Such is our worrisome penchant for inauthenticity, what the early Heidegger calls *das Man*, "the They." In Walter Benjamin's terms, humanity in the age of mechanical reproduction lost contact with the "aura" of what was formerly sacred. The world itself was

Bones, courtesy of Pixabay

therefore lost, profaned, buried under human production and, increasingly, under human consumption – although now open to the possibility of a true mass consciousness. Indeed, the natural world itself was set up to be consumed as mere object. And an "object" is not really what we'd normally call a home. This was a nature designed, as in the Bible, for men to devour and excrete.

Cue the postmodern condition, roughly the period following the Second World War.

12 The Postmodern Condition

According to a well-known formula devised by French philosopher Jean-François Lyotard, postmodernism is "incredulity toward metanarratives" (1984a, xxiv). Unpacking this claim, Lyotard argues that the "postmodern condition" is part of a post-industrial "delegitimation" of the "grand narratives" of philosophy and politics, claims about knowledge and freedom (31, 37), and claims about the subject-who-knows, the "metasubject" in control of the "universal metalanguage" (34, 43). Just as important is the prevailing attitude about this new condition: postmodernists are not even nostalgic for "the lost narrative[s]" (41). Postmodernists are, on the contrary, celebratory – dancing on the grave of modernity and its rational humanism.

One implication of this diagnosis is that "postmodern philosophy" no longer serves the traditional function of legitimizing knowledge and, in the extreme, of legitimizing politics. Philosophy is therefore a discipline in "crisis" (41). Like "art," in Benjamin's telling, it is profaned. Lyotard means philosophy across both Anglo-American and European traditions, means both philosophy as a handmaiden to science (technocracy) and as a series of efforts to ground knowledge in History (ontotheocracy). The theme of crisis or the "death of philosophy" is therefore a key feature of postmodern scholarship and one of the most controversial claims of Lyotard's "report on knowledge." But when he speaks of "philosophy" he really means the theoretical foundations of Western society more generally and its dominant expression in the institution of the university.

Of course the university, which begins with Plato's Academy and Aristotle's Lyceum, is the institutional expression of the original dream of philosophy, *philo-sophia*, the "love of wisdom." In this sense incredulity toward metanarratives is really an incredulity toward the founding wisdom of Western philosophy, most especially its tautological myths of legitimacy as grounded in the history of philosophy, in a world populated with viable (but mostly ignored and denigrated) alternatives – with "Others." In short, we are still battling the influence of global trade, and with it the world-weary, relativistic, smug know-nothingness that infected cosmopolitan Athens in Plato's own time. Today we are back to where Western thought began, namely, with *philodoxia* or the love of opinion.

As traditional forms of knowledge legitimation falter by the mid- to late twentieth century, the void is filled by naked performance, power, efficiency, and finance (46). In the absence of philosophy – nihilism. And if philosophy fails because belief in the ideals of knowledge and freedom have failed, then the university also loses its overarching *raison d'être*. And so it has. The university today has become a plaything of political and economic agendas, a factory to hone practical skills and create thoughtless functionaries for the benefit of the state (48). The question that increasingly dominates life, Lyotard therefore complains, "is no longer 'Is it true?' but 'What use is it?'" (51). Utility trumps wisdom. The effect is what Lyotard calls the "mercantilization of knowledge." And so it has come to pass, yet again, that Heidegger's fears are realized: mere calculative thinking has "come to be accepted and practiced *as the only* way of thinking" (1955, 56).

Commissioned in 1979 by the Council of Universities of the Quebec provincial government, *The Postmodern Condition: A Report on Knowledge* points to a major cultural and intellectual shift against the norms of Enlightenment thinking. And while it's not Lyotard's best work, it's nonetheless prescient in its estimation of the role of "knowledge in the computerized society," the demise of the university as a place set apart from utilitarian calculation, and the rise of "neo-liberalism" – the ceaseless reduction of life to "mercantilization." More about all of this in section 2.

13 "Knowledge is matter for TV games" (Lyotard 1984a, 76)

In a follow-up to *The Postmodern Condition*, Lyotard returns in 1982 with an essay called "Answering the Question: What Is Postmodernism?" – a key nodal point in the literature after Kant and a pointed response to his critics and to critics of postmodernity more generally. In it Lyotard surveys what he considers to be a retrograde reaction against a postmodernity associated

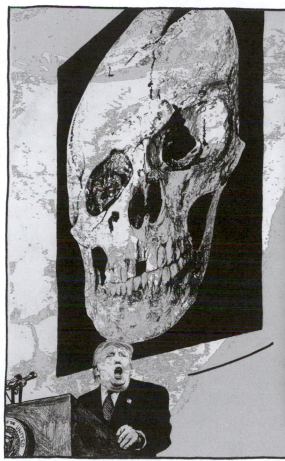

"Smile" (2017), Dwayne Booth

"Rethinking Consumer Capitalism" (2018), Andy Singer

with difference, fragmentation, and disorder. Hence his opening remark, not about postmodernism but about a creeping nostalgia for unity, totality, and order: "This is a period," begins Lyotard, "of slackening" (1984b, 71). What has slackened, just to be clear (because Lyotard isn't), is our commitment to postmodernism.

The word "slackening" appears a second time half-way through the essay, where the rise of eclecticism in art and a corresponding "absence of aesthetic criteria" has created a meaning vacuum filled by commerce. "Artists, gallery owners, critics, and public wallow together in the 'anything goes,'" Lyotard writes, "and the epoch is one of slackening" (76). Lyotard immediately connects slackening to a newfound "realism," about which he adds, "But this realism of the 'anything goes' is in fact that of money." Ultimately it is neoliberal capitalism, not postmodernism, that *gives value* and thereby stamps meaning on reality. For Lyotard, postmodernism is always on the side of experimentation, difference, and the critique of reality – always, it's safe to say, on the side of the angels.

The theme of slackening returns once more, this time in Lyotard's conclusion, where he writes, "Under the general demand for slackening and for appeasement, we can hear the mutterings of the desire for a return of terror, for the realization of the fantasy to seize reality" (82; cf. 64). For Lyotard, postmodernism shares with high modernism a commitment to what is unrepresentable, to what is beyond representation, beyond "reality" – not memorialized, as he will say a few years later, but "immemorialized." His essay takes this commitment literally, spending a lot of effort thinking about avant-garde art and aesthetics more generally. An example, mentioned in passing, is "White on White" (1918), a nonrepresentational composition in white painted by the Polish Russian artist Kazimir Malevitch (78). Lyotard's target in this regard is Kant's notion of the sublime, "ideas of which no presentation is possible" (78). According to Lyotard, Malevitch's all-white painting succeeds to the extent that it evokes the sublime beyond realist representation and thus poses questions that are essentially conceptual – that is, philosophical.

Lyotard's interest in Kant's unrepresentable sublime spawned a decade of interest among readers, especially in English departments, interested in the revolutionary potential of art to subvert (official and officious) representation. But Lyotard's reach was more ambitious than representation in and through art, and extended far beyond his reflections on the postmodern moment. For example, in *Heidegger and "the Jews"* of 1988 Lyotard speaks more pointedly about how "the immemorial" best captures the impossibility of testifying to or representing the trauma (cast in Freudian language) of the Holocaust. The stakes really couldn't be higher than the stakes of life and death, and the attention given to aesthetics, however "radical" and political, has sometimes obscured this fact.

For Lyotard, one memorializes the past, most especially the traumas of the past, at the high cost of violating the past, of "terrorizing" what cannot and should not be reduced to a given representation. And so, if postmodernism is incredulity toward metanarratives, it's because "The nineteenth and twentieth centuries have given us as much terror as we can take" (81). Postmodernism is therefore a *justifiable* response – aesthetically, politically, philosophically, and above all ethically – to those grand narratives that unify and order, but by the same token gather up and liquidate. Enlightenment kills difference. It is perhaps not too absurd to say that postmodernism is indeed, for this reason, a *realistic* response to the "terror" of Enlightenment – a fresh way of characterizing a suspicion about the reign of reason that crystalized with Adorno, Marcuse, and the Frankfurt School and that runs through the ethical reflections on the Other as attempted by thinkers like Emmanuel Levinas.

The last sentence of Lyotard's essay abruptly announces his answer to the question of postmodernism. He writes, "The answer is: Let us wage a war on totality; let us be witnesses to the unrepresentable; let us activate the differences and save the honor of the name" (82). Note Lyotard's prescriptive language, his call to action: "let us wage, witness, and activate." The final answer to the question of postmodernism is that we should "save the honor of the name" of postmodernism. For the "is" of realism is on the side of the powerful, while the "should" of postmodernism is on the side of the powerless. In other words, the answer to the question of postmodernism is a moral imperative to immemorialize the Other.

Briefly compare this conclusion with Kantian ethics. According to Kant's celebrated maxim, the categorical imperative, we should "Act only according to that maxim by which you can at the same time will that it should become a universal law." Lyotard's postmodern imperative is essentially an inversion, very nearly a parody, of Kant's maxim: "act only according to the condition that explodes the illusion of universal law." Such is Lyotard's self-described "agonistics" or "parology," the practice of contesting establishment philosophy and reason (16, 60).

14 Opacity

So what does Lyotard want of philosophy and, by extension, of the university? He wants, he says, "reflexive experimentations" (Lyotard 1984a, 193). In yet another essay on postmodernism Lyotard argues that "philosophy heads not toward the unity of meaning or the unity of being, not toward transcendence, but towards multiplicity and the incommesurability of works. A philosophical task doubtlessly exists, which is to reflect according to opacity" (1984c).

Descartes famously sought out "clear and distinct" ideas. By the time of Lyotard, aspirations had shifted massively. A philosopher's job is not to reduce, simplify, clarify, systematize, analyze, and unify but to embrace and even validate the complexity and confusion of things – the opacity of existence. A philosopher, like an artist, thus proceeds by way of experimentation. This, not incidentally, is why the best artists are like philosophers, as they raise questions about meaning, and why the best philosophers are like artists, as they raise questions about creativity and artistry.

What about postmodernism? In a way, we aren't very far from Foucault's commitments. Postmodernism is duty-bound to champion the Others, the conceptual antiheroes, the "queers," excluded by the closed Enlightenment discourses of knowledge and freedom as expressed in traditional philosophy and politics. Postmodernism champions difference and dissensus over sameness and consensus; the partial, incomplete, and uncertain over the cumulative, final, and certain; the sublime and perhaps even sacred experimentation (in art and elsewhere) over pedestrian and officious realism.

It's useful to note that Lyotard's defence of his report on postmodernism is actually quite personal. For in the end he defends a form of French poststructuralism against the grand narratives of the emancipated and unified community advanced by a fellow critic, the German social philosopher Jürgen Habermas. For it is Habermas's appeal to community that, by Lyotard's reckoning, represents a retrograde "slackening" in the thrall of "terror" and "appeasement" – two incredibly loaded words to fling at a well-intentioned critic.

15 Presentism

Lyotard's definition of postmodernism has the considerable merit of focusing critical attention not on transient fads and fashions but on truth, meaning, and the rational subject as codified in the era of Kant's Enlightenment. In this sense Lyotard answers questions about his present just as surely as Kant did, albeit without Kant's commitment to a certain kind of philosophy, a certain kind of subject, and a certain kind of comportment toward the future. In fact Lyotard's use of Kant's third *Critique* is highly selective – or discriminating – appealing to judgment and art, the realm of feeling, without endorsing Kant's own corresponding belief in knowledge, obligation, and community.

And actually Lyotard's answer to the question "What Is Postmodernism?" forecloses entirely on the possibilities of the future. In this respect his response to postmodernism is, fundamentally, very much a French affair. After the disappointment of Soviet communism and subsequent crises of confidence among French Marxists in the 1950s, and after the depressingly ineffectual student protests of 1968, the future had lost its charm. There may be a beach beneath the cobble streets of Paris, as the Situationists declared as a rallying cry for change, but there were always far nicer beaches elsewhere. Like post-Napoleonic Europe, the disenchanted present trumped all consideration of an enchanting future. Hence the retreat, among French intellectuals, to "revolution" at the level of individual psyche – the retreat, in short, to psychoanalysis. Hence the career of Jacques Lacan and with him a different kind of call for change: *changer la vie*, change your life.

MAY 1968

In the wake of May '68 and its utopian dream, there weren't that many exciting intellectual projects around. May '68 had failed, you couldn't believe in Marxism anymore, and Structuralism was hopelessly removed from life. Psychoanalysis, with its transgressive and initiatory aspects, seemed to be the only theory left that could claim to effectively "change life" – *changer la vie*, as the May '68 slogan would have it. (Mikkel Borch-Jacobsen 1994)

But unlike post-Napoleonic Europe, which was characterized by late romantic pessimism, a bummer philosophy, postmodernity was filled with revelry and celebration. Hence the revival of interest in Nietzsche's commitment to Dionysian revelry, play, laughter, and dancing. Hence the Sex Pistols – with a hint of nihilism ("There's no future, no future, no future for you") thrown in to heighten the frisson of danger.

Like many of his French contemporaries, Lyotard could no longer believe in the future, Marxist or otherwise. He therefore accepted, fatalistically even, capitalism as a given metacondition that warranted resistance but not revolution (see William 2000, 29). On this reading, postmodernism is effectively a form of poking the bear. So yes, there is in the later Lyotard a limited appeal to resistance: "let us wage a war on totality." But he is once again not far from Foucault's own belief that we must nonetheless abandon the "dream of revolutionary change."

Of course, such fatalism is precisely why Habermas rejects postmodernism as unwittingly allied with neoconservatism. For postmodernism very clearly renders thinking about collectivities, and about the future, not just naïve but impossible. The difficulty for readers is that "neoconservatism" is essentially the same charge Lyotard makes when he invokes the "slackening," "appeasement," and "terror" of Habermas's philosophy of community and consensus. One is therefore caught between two well-intending and, in their own ways, convincing arguments about the conservative limitations of the other thinker's arguments.

16 Saving Nietzsche

Let's stick with Lyotard. One problem is his fairly narrow definition of postmodernism. For it's obvious that postmodernism (practically as a matter of principle) subsumes many different, often contradictory, ideas and practices beyond "incredulity toward metanarratives." Perhaps the best laundry list of postmodernisms is assembled by Dick Hebdige in *Hiding in the Light* (1988), where he includes everything from "the décor of a room, the design of a building, the diegesis of a film" to the "anti-teleological tendency within epistemology, the attack on the 'metaphysics of presence'" to the "disillusioned" baby boom generation and more (181–2). Suffice to say, with Hebdige, that we are "in the presence of a buzzword" – namely, a word that has come to mean anything and nothing at all.

Hebdige's genealogy is more useful. He argues that postmodernism was determined, on the one hand, by the depressive aftermath of the May 1968 student uprising and, on the other, by the ongoing attempts (especially in France) to save Nietzsche from Heidegger's reductionist interpretation that appeared only in the 1960s. The result, he claims, is "three negations" at the heart of postmodernism: a movement *against* totalization, *against* teleology, and *against* utopia (186–203).

Hegel and Marx are the preeminent boogiemen of all three negations, and Nietzsche's influence cannot be underestimated. In fact all three negations can all be traced back to Nietzsche's famous, pithy observation (from *The Gay Science* of 1882) that "God is dead." For if there is no secure, divine measure of all things, then the ideal of objective, universal, and transcendent "Truth" is also dead – along with any philosophy that hooks its wagon to that project. Of course for Nietzsche the "death of philosophy" isn't especially regrettable, since the Enlightenment project to know truth, built on the back of an ontotheological belief in a transcendent God, is the very definition of a life-negating, and also delusional, form of nihilism. This is why, in Nietzsche's formula, metaphysics *is* nihilism. And this is why Heidegger, the advocate of Being (ontology, existence) over metaphysics (essence), was attracted to Nietzsche in his middle and late phases.

ON "GOD IS DEAD"

How to understand our cheerfulness. – The greatest recent event – that "God is dead"; that the belief in the Christian God has become unbelievable – is already starting to cast its first shadow over Europe. To those few at least whose eyes – or the *suspicion* in whose

eyes is strong and subtle enough for this spectacle, some kind of sun seems to have set; some old deep trust turned into doubt: to them, our world must appear more autumnal, more mistrustful, stranger, "older." But in the main one might say: for many people's power of comprehension, the event is itself far too great, distant, and out of the way even for its tidings to be thought of as having arrived yet.
(Friedrich Nietzsche 1882, 199)

After Marx, Nietzsche was the thinker most committed to diagnosing his present time as a harbinger of a future yet to come; a future linked to the "overman" and the will to power, the very ideas that influenced Hitler and Nazi ideology, as well as the egoistic fantasy lives of countless teenage boys ever since. French thinkers in the 1960s were trying to save Nietzsche, but also Western culture, from this historic misappropriation by Hitler (and by other misunderstood boys). But they were also trying to save Nietzsche from Heidegger's too masterful, too totalizing interpretation and appropriation. An unsentimental postmodernism was just the kind of Dionysian overcoming that a "new Nietzsche" countenanced: a future without sentimental nostalgia, sceptical of rationalistic philosophy, contemptuous of absolutist and universal measures (God, Truth) of human existence, and disdainful of all forms of essentialism, including the hermeneutic reduction of a thinker's messy claims to One Big Thought (for example, the reduction of Nietzsche's incredibly open thinking to the thought of the will to power). So down with anti-Semitism, right wing fascism, left wing totalitarianism, nationalism, Nazi concentration camps, Soviet gulags and psycho-prisons, capitalism, communism, colonialism, and imperialism. And down with the grandiose projects to organize and finally know the world-in-itself, whether Hegelian, Marxist, Freudian, or Structuralist. Down with any project, based on an initiate's knowledge of occult depths, that promises grand demystification, utopia realized, and the cure for the problems of everyday existence.

In short, the intellectual Left of the late 1960s and 1970s finally had its own way with a Nietzsche understood as absolutely resistant to the charms of systematization, ultimately to the charms of Nazism. This is not quite the "poet-Nietzsche" of old. This is Nietzsche as critic – the philosopher of the *end* of philosophy. And these are the roots of French "poststructuralist" philosophy.

Poststructuralism amounts, in large part, to a historical demolition of the rational subject that occupies Kant – and which underwrites everything that goes wrong after Enlightenment. Poststructuralism is in this respect sometimes called a form of "antihumanism" and is, as such, thoroughly Nietzschean in theory and sentiment. As for postmodernism, the dumb cousin of the far more philosophically sophisticated poststructuralism, it's not always aware of this problematic history of the "universal subject" of reason. Why is that? Because the philosophy is often buried under the cultural detritus of fad and fashion as expertly recounted by Hebdige and other cultural critics. In this respect philosophy, still a haughty product of high culture, has in the twentieth century been eclipsed by journalism and sociology, the study of "low" or mass culture. This is why Lyotard, still very much a philosopher's philosopher, is unapologetically interested in high modernism, the avant-garde, and artistic experimentation. As for mass culture – practically the calling card of the postmodern era – he leaves that to others, for example, to Jean Baudrillard, the sociologists, and (perhaps most especially) to the English and comparative literature professors.

Yet the continuing haughtiness of philosophy, including Lyotard's "philosophy of opacity," is very obviously a problem. For it's a philosophy insufficiently attuned to the *masses* and to *culture*; insufficiently attuned, therefore, to our era. If philosophy can ever hope to diagnose mass society and the mass subject of the present, then it learns very little from modernist shibboleths about art, aesthetics, and upper class snobbery. Put otherwise, if God is dead then we should stop looking to the church (museum, gallery, university) for all our answers. The truth is that Lyotard, ill at ease with the very postmodernity he champions, rationalizes his present by carefully cherry-picking the past – not unlike the curator of a highbrow museum.

My own view, therefore, is this: philosophy today must contend more radically, not with the "history of the present" as examined by Foucault and Lyotard but with the *future of the present*. And the future of Lyotard's and Foucault's postmodern condition is, quite simply, the moment that I am calling the Anthropocene condition or anthropocenity. As for philosophy, it must reenchant humanity with the future and not simply disenchant us with the present-as-determined-by-the-past. The only question is what can we learn about our possible futures from a history that culminates, most recently, with postmodernism.

17 Society without God

American pragmatist philosopher Richard Rorty, among the most clear-eyed and ecumenical thinkers in recent memory, weighs in on the debate between Lyotard and Habermas on the question of postmodernism, attempting, he says, "to split the difference" (1984, 42). In "Habermas and Lyotard on Post-Modernity" (1984), Rorty registers his disagreement over the general direction of philosophy across both Anglo and European traditions, arguing that the long interest in the "philosophy of the subject," starting with Descartes, has been a mistake. According to Rorty, we'd all be better off had we followed Francis Bacon – "the prophet of self-assertion, as opposed to self-grounding" (39). Which is fair enough. But the truth is, it's *our* mistake, and tales of "what if" don't really touch it – or matter.

Rorty's candid conclusions about how to respond to Lyotard and Habermas are more useful. Influenced by American pragmatist John Dewey, Rorty favours reenchantment with a world of "the concrete" (42). "This Deweyan attempt to make concrete concerns with the daily problems of one's community," argues Rorty, "seems to me to embody Lyotard's postmodernist 'incredulity toward metanarratives' while dispensing with the assumption that the intellectual has a mission to be avant-garde, to escape the rules and practices and institutions which have been transmitted to him in favor of something which will make possible 'authentic criticism'" (42). For Rorty, Lyotard's turn to the sublime and immemorial is in fact "one of the prettier unforced blue flowers of bourgeois culture." As such it is "wildly irrelevant" to achieving the kind of vital consensus associated with Habermas; is indeed a "hopeless attempt to make the special needs of the intellectual and the social needs of the community coincide" (43).

If Rorty splits the difference between Lyotard and Habermas, it's because he appreciates the intellectual and social reasons for thinking about his postmodern present. Yet it's important to register, with Rorty, that both sides seek to bring about the end of philosophy. The philosophy of opacity, inspired by Nietzsche, ends with avant-garde experimentation and play. The philosophy of community ends with concrete solidarity and freedom. "Those who want sublimity," Rorty concludes, "are aiming at a postmodernist form of intellectual life. Those who want beautiful social harmonies want a postmodernist form of social life, in which society as a whole asserts itself without bothering to ground itself" (43).

It is a staple of postmodernist thought that foundations have crumbled, that philosophy (as metaphysics) has come to some kind of end, that legitimacy is just the naked by-product of performative success or, in the language of Ludwig Wittgenstein, a given language game or, in Rorty's terms, an interpretive community. Society "asserts itself," Rorty says, "without bothering to ground itself." Once more, both the intellectual and social worldviews that Rorty characterizes are in this sense heirs to Nietzsche's demolition of universal Truth. Postmodernism is Nietzschean, the destruction of false idols. The task of philosophy therefore shifts from legitimizing God and Truth to explaining the unstructured, ungrounded communities that characterize the postmodern condition: to explaining society without God.

But that's the postmodern condition. Obviously the Anthropocene condition is under no obligation to continue in this tradition of counter-Enlightenment thought, from late Romanticism to postmodernism, or to simply reaffirm, in a futile and retrograde state of nostalgia, the ideals of Enlightenment. The Anthropocene condition is free to ground its claims in whatever ways suit a present turned toward the future – perhaps even in Rorty's terms as ungrounded self-assertion, reenchantment with the concrete, and "untheoretical social solidarity" (40–1). Even so, it probably shouldn't do so blindly and ahistorically, either. Confronting our intellectual history is one way of confronting the present while laying the ground for thinking the future. Yet it also can't be the last word.

18 Summary

Before we shift from a retrospective analysis of philosophy to the conditions of anthropocenity – to the present – let's quickly summarize the territory just covered.

Kant basically resolves the conflict between the opposing epistemologies of rationalism and empiricism, innateness and experience – the two classic ways of explaining how we know what we know about the external world. But the resulting critical philosophy leaves us with a new problem: there is a veil between subjects and the object world, between the phenomenal realm and "noumenal" realm of the *thing-in-itself, das Ding an Sich*. All knowledge is a human and mediated knowledge, to wit, is a represented knowledge.

Hegel thinks he resolves this problem by finding in History itself all the component parts that make up our understanding of the object world. Conflicts between ideas are resolved and perfected over History, and the realization of this process in the wake of Kant's great synthesis is called absolute knowledge.

However, Marx argues that Hegel's solution is at best an intellectual solution, one still disconnected from the external world it claims to know. So although Hegel hits upon the right methodology, dialectics, his focus on ideas and consciousness is itself a symptom of false consciousness. Marx turns instead to what he considers the truth of History, namely conflict at the level of material production, economics. Human beings reshape the world and are reshaped in turn by that world. That is his basic answer to the riddle of how human beings know the object or external world: we know the world to the very extent that we reshape it (and it reshapes us) through our labour on it. Or again, we know this world because our labouring beings coincide with it. As a consequence, the problem of knowing the world by *representing* it goes away, replaced by the achievement of a human freedom won through meaningful human activity in, on, and to the world. In a way, freedom is just this overlap of the labouring subject with her world. One *is* because one *does*. Identity is work.

Neo-Marxists like Marcuse noticed, however, that our fashioning of the world had only refashioned a perverted identity. Instead of an authentic being-in-the-world with Heidegger and instead of collective freedom with Marx, subjects in the twentieth century had become reflections of capitalism: the wage earner cut off from Being because cut off from meaningful labour, and ultimately fashioned as a consumer. The neo-Marxist critique of capitalism therefore shifts attention from capitalism proper, from economics, and toward the subjects of capitalism; away from the ideals of a future enlightenment and toward its perversion in a presentist culture of instrumental reason – to the so-called "superstructure" (hence "cultural studies"). This view is a mainstay of Frankfurt School philosophy of society, the echoes of which reverberate straight through to our most recent postmodernity.

In short, the perverted psyche under capitalism is not the Enlightenment subject, a universal subject of reason, but the desiring and distracted subject of irrationality. This distracted subject, first theorized by Benjamin, is one powerful way of thinking about the postmodern subject. It is the subject given over to advertising and marketing. One is "free" to work and to shop.

19 A Differential Consciousness?

In our own lifetime, however, we have become aware that the background is no longer just a background. We are part of it ... (Dipesh Chakrabarty 2015, 179)

So what can this history tell us about our present moment? What have we learned? First, it's clear that philosophy has never really solved the old problem of *knowing* the external world or, with Marx, of *being* the external world. Our knowledge still remains a highly mediated knowledge. Second, the experience of crisis after crisis, of serial black swan events, of economic and environmental collapse, of preventable and widespread death, of the perceived end of our present (economic) conditions, all portend an alteration of human subjectivity, of consciousness, of being-in-the-world.

And so I pose a tentative hypothesis disguised as a question: Could it be, in the Anthropocene condition, that we finally do *know* the world because we *feel* it, feel the climate change, feel the thunder, wind, rain, snow, the noumenal having become concretely phenomenal? In short, can we feel the difference, feel the shape of this *differential consciousness*? Of a subject newly bewildered by an external world – Chakrabarty's "recognition of the otherness of the planet" (2015, 181) – that it has, heretofore, evolved within and, as such, has taken for granted and rendered invisible? Of a subject finally startled by a new awareness, not of the Holocene environment of old, and not even of loss – but of a strangely hostile environment that deserves its own name, the Anthropocene?

Factory, courtesy of Pixabay

DIFFERENTIAL DIAGNOSIS (NOUN)

the distinguishing of a disease or condition from others presenting similar symptoms [medical].
(Merriam-Webster)

This is not quite a hypothesis about Nature or History, about inequality or class, or about dupedness or false consciousness, either. It is a hypothesis about a dawning recognition of the Earth on individual and collective levels. About an Earth that communicates to us as it bakes, fries, freezes, floods, and kills us. About a force called "Gaia." This is not a shift back *to* a given nature, therefore, but a return *of* natural being-in-the-world. It is not the recognition of another subject, as in Hegel, but the utter novelty of recognizing the living force of the Earth (or the "biospheric") as an object that has always responded, by feedback, to human activity. If so, and here is my basic claim, then the changing Earth will bring with it a radical alteration of human consciousness and a renewed debate about what it means to be a knowing subject, a human being after the disaster that has been humanism. It should alter, in turn, what it means to be an "object" in the natural world, from trees to dogs to the Earth itself. If so, then our most recent experience of the postmodern condition was less prelude to the future than postscript to the Enlightenment.

It's over. Let's turn now to consider our "Present."

PART TWO

THE PRESENT (CA 1968–2008), WORLD AS OBJECT OR THE ANTHROPOCENE CONDITION

The state itself, which once represented the interests of the people, even if those interests were often thwarted by the power of business, has been reshaped since the 1970s to serve the interests of the Economy. (Clive Hamilton 2010, 49)

1 Boomernomics

Although the "postmodernism condition" nearly gets us up to speed on changes to the human subject roughly two centuries after Enlightenment, it doesn't entirely help us prepare for, let alone imagine, the subject of the present and near future. After the reign of reason – a deluge of power. After the terror of grand narratives – a plethora of micronarratives. After humanist triumphalism – a celebration of antiestablishment antiheroes espousing antihumanism.

The American English professor Fredric Jameson probably the best guide to the time and structure of postmodernity, equates postmodernism with the post-Fordist, post–World War period of reconstruction, deficit spending, and Keynesian-inspired New Deal megaprojects. Jameson's loaded term for postmodernity is therefore "late capitalism" (a term that goes back decades). It's not surprising that Jameson, as a Marxist, looks to the economic base to explain the superstructural ephemera of postmodern culture or that he lends special credence to the time of capitalism, the stage (according to the argument) that makes possible a future communist society. Marxists are always on the trail of deep (economic) structure and are dogmatic about the radical potential of the era defined by capitalism. That's also what made Marxists so out of step with the postmodern celebration of disunity and corresponding disregard for the future but very much in step with exposing the excesses of late capitalism. Just as the Anthropocene is good for philosophy, late capitalism has been good for Marx and Marxism.

But one doesn't need to be a Marxist to theorize the importance of "the economy" – long a pleasant, if obscuring, euphemism for the myriad forms of actually existing forms of capitalism – to understanding contemporary existence. It was Bill Clinton's campaign, during the 1992 presidential race, that formulated the well-known refrain, "it's the economy, stupid." And it

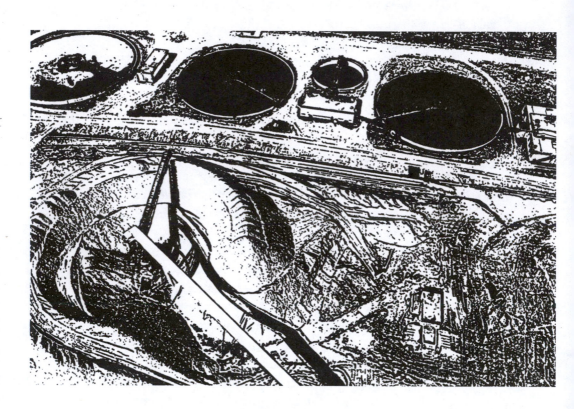

"Bowater" (2017), Mark Nisenholt

was under that banner that Clinton, and following after him Britain's Tony Blair, pushed through the neoliberal policies of deregulation that generated impressive wealth at the turn of the century and, in 2008, caused such a devastating and interminable global depression.

Since the 1990s scholars and journalists have employed a constellation of terms to describe this new era. Here is feminist social theorist Clara Sacchetti's laundry list of overlapping terms: "postmodern capitalism, contemporary capitalism, recent capitalism, cannibal capitalism, crack capitalism, stakeholder capitalism, natural capitalism, coordinated, state, or free capitalism, flexicurity capitalism, extreme capitalism, global capitalism, informational capitalism, finance capitalism, late capitalism, consumer capitalism, and neoliberal capitalism" (2013, 4).

To this list of terms we can add another, this one inspired by the dominant sociological force that made all this change possible, namely the American baby boom generation of Bill Clinton – the true heir of Reagan. Let's call it *boomernomics*, the meaning of which is broad but arguably significant: the shift from production and need to consumption and desire; the high value assigned to audit culture and metrics; the quantification of job performance; the belief that efficiency and continuous improvement strategies are good in themselves; the reverence given to strategic planning, mission statements, outcomes, and managerial expertise, essentially as legitimized by the MBA degree; the dogma of tax reductions, hatred of "big government," and the de facto institutionalization of trickle-down economics; the ratification in practice and in law, by the United States Supreme Court, of corporations as legal persons; the campaign finance manipulations of democracy, largely through corporate "free speech" (e.g., American "Super PACs"), in the service of the extremely wealthy; the signing of complex free-trade agreements that explicitly place corporate profits ahead of collective national interests, including social security, health care, welfare, energy, transportation, and water programs; the privatization and, consequently, destruction of public institutions and public spaces around the world, as driven, for example, by powerful lending organizations like the International Monetary Fund; the deregulation of financial and environmental standards leading to worldwide economic and environmental crises and, finally, catastrophes; constant recourse to debt financing, effectively stealing money from future generations; the perilous rise of financialization, a fantastical form of extracting wealth from the poor; the wilful deindustrialization of the West and, with it, the shift of manufacturing jobs to regions of the world with the lowest wages and the weakest environmental and labour regulations; the subsequent destruction of "First World" labour unions and, as a consequence, the strangulation of middle class existence everywhere; the prevalence of austerity as the preferred strategy for governing and/or enslaving nations, beginning in the Global South but ultimately utilized as well in the Western world (e.g., Greece); the jarring spread of "sacrifice zones," those places of major resource extraction and environmental destruction, from "hidden" locations in the developing world to the backyards of the developed nations (e.g., through "fracking"); the monopolization of information, and with it the demise of investigative journalism and informed public debate; the gutting of public universities, including the firing of professors, the abandonment of tenure (e.g., the United Kingdom and Australia), and the quashing of dissent everywhere; the delegitimizing and dismantling of the critical arts, like philosophy, and the defunding and/or muzzling of scientific research, especially environmental research; and so on *ad nauseam*.

THE TELEGRAPH 2012 HEADLINE

"Mass suicide" protest at Apple manufacturer Foxconn factory

Around 150 Chinese workers at Foxconn, the world's largest electronics manufacturer, threatened to commit suicide by leaping from their factory roof in protest at their working conditions. (Moore 2012)

Gas station, courtesy of Pixabay

SACRIFICE ZONES

It's absolutely imperative that we begin to understand what unfettered, unregulated capitalism does – the violence of that system. These [towns in America] are sacrifice zones, areas that have been destroyed for quarterly profit. And we're talking about environmentally destroyed, community's destroyed, human beings destroyed, families destroyed. And because there are no impediments left, these sacrifice zones are just going to spread outward. (Chris Hedges to Bill Moyers, 20 July 2012, see Hedges 2012)

Bluntly put, the generation that advanced the most beautiful ideals of equality and peace quickly devolved into the most efficient, most regrettable, and yet most diffuse force of planetary destruction in the history of humankind.

2 Wither Marxism?

Given the critical importance of American baby boomers as a collective sociological force in the contemporary world, the focus on the *structural* causes of environmental destruction (i.e., capitalism), while of critical importance, is not enough. Which means the strict Marxist interpretation of the environmental crisis is not enough.

One problem is that Marxist recourse to a mechanistically determinative structure can be used to excuse the criminal actions of individuals, blinding us to the need for justice and ethics – for punishment of criminal actors, most especially white-collar criminals. This problem litters Mark Fisher's influential *Capitalist Realism* (2009), where he observes that even individuals with good intentions are corrupted by the bureaucratic structures of capitalism that define their jobs. Consequently, he claims, it's "a mistake to rush to impose the individual ethical responsibility that the corporate structure deflects. This is the temptation of the ethical which, as Žižek has argued, the capitalist system is using in order to protect itself in the wake of the credit crisis – the blame will be put on supposedly pathological individuals, those 'abusing the system,' rather than on the system itself" (69).

In many ways Fisher is obviously right. Power corrupts. And since capitalism is the source of that power, then it's capitalism that corrupts individuals. Logical conclusion: capitalism is responsible for the individual criminality that arises. Investment bankers as the highly remunerated dupes of capitalism. It's on the basis of this structural analysis, furthermore, that Fisher spins out amazingly reductionist arguments about the consequences of capitalism. Take his discussion of mental health. While it's no doubt true that capitalism can and does make people physically and mentally ill, Fisher contends that disorders such as dyslexia and attention deficit disorder are literally caused by capitalism (2009, 25–6), a contention that, despite its pedigree in Marxist-inspired antipsychiatry of the 1960s, is laughably absurd. To be fair, Fisher probably doesn't care. His stated motivations are narrowly political: "the task of repoliticizing mental illness is an urgent one if the left wants to challenge capitalist realism" (37). Fine. But using mental illness as a political football is a dangerous and probably unwise strategy.

Fisher argues that the "impasse" between capitalist structure and individual responsibility is really a "dissimulation"; one that "indicates what is lacking in capitalism." Here's his claim:

> What agencies are capable of regulating and controlling impersonal structures? How is it possible to chastise a corporate structure? Yes, corporations can legally be treated as individuals – but the problem is that corporations, whilst certainly entities, are not like individual humans, and any analogy between punishing corporations and punishing individuals will therefore necessarily be poor. And it is not as if corporations are the deep-level agents behind everything; they are themselves constrained by/expressions of the ultimate cause-that-is-not-a-subject: Capital. (69–70)

For Fisher the "ultimate cause-that-is-not-a-subject" is the proper object of attention and of punishment. Individuals and even corporations are just reflections of this "deep-level" agency based on Capital (rendered in caps). But Fisher too easily separates the subjects and objects of capitalism. According to Marx's theory of "species being," discussed in section 1, the capitalist subject is a *realistic* reflection of the economic base; she helps make, and is in turn made by, capitalism. The historic alignment of subject and object (of subject and her world) is what Marxism discovers in the world. Better stated, Marx finds that one identifies with what one *does* in the world. This action on the side of the subject, the actor, is obviously essential, since the relationship between subject and object world happens on *two sides at once*; that's what we mean when we say it is "dialectical." Consequently it doesn't make sense to try, willy-nilly, to excuse actors from the objective content and expression of their own identities or to reduce them to mere symptoms. It's literally unrealistic.

As for Fisher's question about which agencies are equal to the task of moderating and regulating the impersonal structure of capitalism – well, the answer is clear. These agencies are called governments. The eighty-year gestation of neoliberalism, as wedded to conservatism and religious fundamentalism, may have delegitimized an institution that once balanced the "impasse" of Capital and individual, but that delegitimation was a blunder. Governments (and its many functionaries) *should* enforce laws and values, and it is bootless to shrug this off with a knowing nod to Žižek and "the temptation of the ethical." True, it's unfortunate that such a temptation has been blunted under the conditions of capitalism. And, sure, pointing to "pathological" or "rogue" individuals can be used by capitalism (let's run with this abstraction and not speak more concretely of "capitalists," that is, of the *agents* of capitalism) to deflect responsibility away from its own machinations – from the system. We know very well that capitalism thrives on sociopathic behaviour, that capitalism is always rogue capitalism, always in excess of itself. Ok. But this says nothing at all about the basic correctness of ethics as an ideal – even business or corporate ethics. To suggest otherwise and simply excoriate recourse to *the ethical* is not only too dismissive; it's also a missed opportunity to grapple with everyday ("Main Street") justice and the sort of responsibility that Dale Jamieson highlighted back in 1992. For, as Jamieson argues, this sort of responsibility also represents an important challenge to capitalism. More about it later.

One upshot is that Fisher's numerous but always fleeting discussions of popular culture – movies and rock and roll – are a bait-and-switch

"Regulation" (2012), Andy Singer

tactic, since in the end he always returns to classical questions of economic structure, of capitalism, understood as the "ultimate cause" of all the chaos afflicting our world today. In my view this is not just a dated, annoying, humourless, boring, and reductionist project, as though every new movie or pop song is secretly an expression, however repressed or latent or unconscious, of capitalism. It's also a mistake. Individuals who identify with or embody the (however regrettable) ideals of capitalism are culpable as individuals. Capitalist subjects, no less than corporations, must answer for what the Pope calls their "cheerful recklessness" (Pope Francis, #59 2015). They are as much a generative as constituted force of the dialectic.

Beyond that, and in any case, we are in no way obliged to accept the classical Marxist interpretation that reduces human subjectivity to psychologically damaged weakness and being-without-agency, to victimhood at the hands of universal structure. The abandonment of agency, as illusion or delusion, is the common mistake, fuelled by remarkable condescension, of both Marxist and postmodernist projects. That is in part why both projects are ethical and practical dead ends. For if subjects can *resist*, then they can also *subsist*. If subjects can (at times, even fleetingly) rise above the material conditions of their given (economic) reality, then they can also be active collaborators in a system of structural inequality. False consciousness, dupedness, and victimhood can't get the last word on what it means to be a "subject of capitalism." And it very obviously doesn't, otherwise there could be no "outsider" criticism of capitalism at all. Including no Marxism.

Despite, therefore, Marxism's usefulness as the best available critique of capitalism, including late capitalism, it behooves us to think differently about the future. That's saying too little. Although the demise of capitalism has earned the gloating of Marxists, who have in some ways been proven right, Marxism doesn't get the last word. On the contrary. The demise of capitalism in our time implies, almost by definition, the demise of its critical other, Marxism. After all, if Jameson is right when he says that "Marxism is the science of capitalism" (1997, 175) and if the field called capitalism is on the verge of disappearing, then, QED, there is no longer any use for a science of capitalism. It thus follows that the communist future as foretold by Marxists belongs to the past or, more precisely, to the future of the past. In the past it certainly served a purpose. But it can't entirely play that role today.

All of which brings us to the fourth feature of the Anthropocene condition: the flaming out of Marxism as a viable guide to the actually-existing-present and its future, to our experiences of life in our time. The task of the present age is to discover new critiques appropriate not to the contradictions of capitalism but to its death in 2008 and interminable zombie-afterlife. For a prophetic philosophy of this new future without capitalism.

3 "Post Truth"

My postwar generation and the boomers who followed – we've lived like kings and queens, and we partied like there's no tomorrow, never worrying about the kind of world we were leaving for our children. Well, the party's over. (David Suzuki 2013)

It's hard to say exactly why things went so terribly wrong with the American baby boomers – and then with the rest of us who followed in their wake. But it's very easy to track the decline of civil society against their collective rise to prominence, so much so that it's insufficient to blame capitalism alone for all that has gone wrong since the 1970s. For it was not just an impersonal economic structure that ushered in the tragic changes enumerated under boomernomics but a very specific, very privileged group of Americans; a group, given its majority status, that determined every major cultural and policy shift as it has moved from childhood to dotage.

In other words, the boomers were not just a reflection of an economic *structure* but were a generational *force* (or, if you insist, a cultural *super-*structure) that must be taken into account – and judged by its actions, not by its sometimes-pretty rhetoric.

The sad truth is that the boomer generation's most memorable characteristic is not advancing civil rights, gay rights, labour rights, and women's rights, as it's often proclaimed. With hindsight its most memorable characteristic is desire for profit-above-all-else, boomernomics, and, by extension, the installation of advertising culture into everyday life. It is, in effect, the saturation of society in the techniques of mass manipulation, sophistry, and lying. For with the ascendancy of boomers, lying and bullshit have become

WHAT IS SO SCREWED UP ABOUT YOUR PARTY IS THAT YOU DON'T KNOW HOW TO MAKE THE COUNTRY FEEL REALLY GOOD ABOUT THE SELF-SERVING, CORPORATE ASS-LICKING, HUMAN RIGHTS-ABUSING, CIVIL LIBERTIES-SMASHING, CLIMATE-KILLING, AMERICAN EXCEPTIONALIZING JINGOISM THAT MAKES US THE GREATEST NATION ON EARTH AND WE DO.

MR. FISH

"Left Right ... Left Right Left" (2016), Dwayne Booth

the new common sense. Its first casualty was truth: a casualty that infected politics and, in turn, undermined evidence-based science. Its second casualty was ethics and care for the Other: a casualty that infected business at every level. Its third casualty? All the hard-won civil rights gains that have been muted, attacked, and gutted since the 1980s when boomers finally assumed almost complete power in every reach of contemporary American society – and well beyond. Philosopher Stephen Gardiner is therefore right to ponder the legacy of boomers: "we may end up being remembered not just as the profligate generation, but as 'the scum of the earth,' the generation that stood by as the earth burned" (Gardiner and Weisbach 2016, 4).

ORIGINS OF ENVIRONMENTAL BULLSHIT

[E]ven the major media refuse to clearly expose
the undermining of real environmental science,
and the creation of lies and bribes to distort public
policymaking. But this work is out there. It's really the
thorough work done on the climate denial machine
that lays out the methodology of the development of
environmental distortions, lies and post-truth discourse.
(David Schlosberg 2017)

It's amazing to realize that this sociological-cum-natural catastrophe is only a few decades old. As journalist David Wallace-Wells put it in an article in 2017, "[M]ore than half of the carbon humanity has exhaled into the atmosphere in its entire history has been emitted in just the past three decades; since the end of World War II, the figure is 85 percent. Which means that, in the length of a single generation, global warming has brought us to the brink of planetary catastrophe, and that the story of the industrial world's kamikaze mission is also the story of a single lifetime" (Wallace-Wells 2017a). That single lifetime overlaps with the lifespan of the boomer generation – or, better stated, ends well before their own children, the "millennials," reached adulthood.

The boomers inherited a rich, dynamic country and have gradually bankrupted it. They habitually cut their own taxes and borrow money without any concern for future burdens. They've spent virtually all our money and assets on themselves and in the process have left a financial disaster for their children. (Bruce Gibney, in Illing 2018)

The young boomers giveth, and the old boomers taketh away. I'm afraid this selling-out of beautiful, noble ideals, including a better future for all, is the concrete reality of their legacy. That and a much-accelerated shift to the Anthropocene.

4 Trumpism

We can't spin our way out of environmental disaster, although that has been the tactic of the people most desperate to hang on to power. For of course the super wealthy and their minions, the feckless technocratic and political class, are the biggest liars of all. To be fair, their victims now include most members of the baby boom generation itself. Aside from an elite few, no one has escaped unharmed by this force.

At the dawn of anthropocenity the fog is lifting. Hence the fifth and possibly most disturbing feature of the Anthropocene condition: the almost universal disgust, very much earned, that the masses feel for politicians and for democracy more generally; a disgust that presents as populist rage, less on the impotent Left than on the lunatic Right. As Chris Hedges puts it, "Trump has tapped into the hatred that huge segments of the American public have for a political and economic system that has betrayed them. He may be inept, degenerate, dishonest and a narcissist, but he adeptly ridicules the system they despise" (2018). The masses simply presume, with remarkable cynicism, that all politicians will lie, cheat, and ultimately screw them. At least Trump, in the process, echoes their own contempt for what is in fact contemptuous. And so it has happened that, by the end of the postmodern condition, the most significant accomplishment that boomernomics has "turned on, tuned in, and dropped out" is the incipient

Recycled plastics, courtesy of Pixabay

bullshit of contemporary fascism. Trumpism: the last tragicomic gasp, or implosion, of boomernomics as politics.

Marx and Engels put it perfectly in *The Communist Manifesto* more than 150 years ago, saying that the bourgeoisie "produces, above all, its own gravediggers" (1848). And so it has. With boomernomics, capitalism has cannibalized itself – has strip-mined the conditions of its own existence – leaving rage and confusion in its place. So here's another more hopeful feature (the sixth) of the Anthropocene condition: with the aging of the baby boom generation we are finally seeing a future beyond greed and bullshit; a future better suited to gravediggers, truth-tellers, and court jesters.

> The system of so-called neoliberal globalisation is not sustainable. It creates a lot of resistance, heroic resistance in the South, and China is also trying to play with it. It has created a huge problem for the people of the US, Japan and also Europe. Therefore, it is not sustainable. Since it is not sustainable, the system is looking to fascism as a response to its growing weakness. That is why fascism has reappeared in the West. It is also exported to our countries. Terrorism in the name of Islam is a form of local fascism. And today you have in India Hinduist reactions. That is also a type of fascism. (Samir Amin et. al. 2018)

BRAND ON REVOLUTION

> We British seem to be a bit embarrassed about revolution, like the passion is uncouth or that some tea might get spilled on our cuffs in the uprising. That revolution is a bit French or worse still American. Well, the alternative is extinction so now might be a good time to re-evaluate. (Russell Brand 2013)

"The Son Also Rises" (2017), Dwayne Booth

5 Backyard Apocalypse & the Democracy of Suffering

The puncturing of boomer self-satisfaction about their innate awesomeness is best registered in Thomas Frank's highly perceptive *The Conquest of Cool: Business Culture, Counterculture, and the Rise of Hip Consumerism* (1997). There he demonstrates that the true legacy of the 1960s was the reactionary neoconservatism of young Republican activists like Patrick Buchanan. The tangible result was the development, on the one hand, of populist talk radio and, on the other, of well-funded and highly partisan "think tanks" – front lines in the low and high cultures of neoconservatism devoted to unseating the policy and cultural gains of peace, love, and understanding (or, if you prefer, sex, drugs, and rock 'n' roll). Hence the rise of Ronald Reagan, Rush Limbaugh, and the culture wars of the 1980s; the Washington-based Heritage Foundation, the Chicago-based Heartland Institute, and the spread of dozens of other neoconservative think tanks as sources of moral and financial support for partisan scholarship and advocacy; reactionary screeds like Allan Bloom's *The Closing of the American Mind* (1987); race-baiting books like Richard Herrnstein and Charles Murray's *The Bell Curve: Intelligence and Class Structure in American Life* (1994) and Philippe Rushton's *Race, Evolution, and Behavior* (1995); economic policy as driven by Milton Friedman and his disciples; political philosophy as driven by Leo Strauss and his disciples at the University of Chicago; Ayn Rand, the "philosophy" of Objectivism, and the dimwitted cult of *Atlas Shrugged*; neoconservative operatives, like Paul Wolfowitz, working within government to provide intellectual cover for Dick Cheney, George W. Bush, Donald Rumsfeld, and their fraudulent war on Iraq; opportunistic centrists like Bill Clinton who championed radical financial deregulation and, as a consequence, is uniquely responsible for the US housing bubble and subsequent collapse of 2008; and of even Barack Obama, who immediately upon being elected president recruited Wall Street insiders into his administration, failed to pursue criminal charges against thieving bankers, and failed to institute meaningful financial reforms to stop it all from happening again. Only blind racism has allowed Americans to believe that Obama, another Ivy League president, was a serious threat to the neoliberal world order.

The widest frame for understanding the consequences of boomernomics is therefore painfully evident. It represents nothing less than a bloodless coup of American government by an assortment of thugs, namely the bloviators, eggheads, flakes, wonks, appeasers, thieves, criminals, liars, and

Junk heap, courtesy of Pixabay

assholes, many strangely connected to Chicago, who collectively and often knowingly paved the way for the highly un-American oligarchy of the present. Enter Donald Trump.

CITIZEN TRUMP 1999

My entire life, I've watched politicians bragging about how poor they are, how they came from nothing, how poor their parents and grand-parents were. And I said to myself, if they can stay so poor for so many generations, maybe this isn't the kind of person we want to be electing to higher office. How smart can they be? They're morons. (Donald Trump, in Dowd 1999)

At the dawn of the Anthropocene, American power has turned decidedly against the American people, so much so that only the most naïve and foolish could mistake their form of institutional terrorism for patriotism.

Naturally the coup has been "bloodless" only in the strict sense – since it's very clear that many people had to die so that boomernomics could live. It used to be the invisible people of the Global South. But in the last few decades the death and privation have spread all the way back to the Global North. Which, let's be honest, is the only reason we're all so concerned about it today. Pope Francis calls this fundamental, ubiquitous disregard for others "the globalization of indifference" (#52 2015). That sounds about right.

THE POPE ON SUFFERING

Sadly, there is widespread indifference to such suffering, which is even now taking place throughout our world. Our lack of response to these tragedies involving our brothers and sisters points to the loss of that sense of responsibility for our fellow men and women upon which all civil society is founded. (Pope Francis #25 2015)

The truth is our rising interest in climate change is clearly driven by enlightened self-interest in the West. In short, and contrary to Naomi Klein's wishful thinking, it's fuelled by old-fashioned NIMBYism. This is all very apparent in Adam Briggle's "philosophical ethnography" of hydraulic fracturing ("fracking") of shale in Denton, Texas. Briggle discovers that the laws are narrowly focused on local land use concerns, while the wider global energy context is treated as an abstraction (see Briggle 2015, 47–9). It's a bias that fits very well into the highly local, personal, even narcissistic interests that citizens have in their own immediate situations – and on the Left and Right alike. So "climate change" isn't on the agenda, only zoning bylaws and "neighborhood integrity." As one small-town mayor puts it, having left his small town of Dish, Texas, for the sake of his children's health, "I used to be a Republican like these people [members of the oil and gas industry], but they have no respect for property rights, which is supposed to be the bedrock of conservatism ... I mean, you look around at what happened in Dish, to my neighbors and family, and you wouldn't believe this is America" (in Briggle 2015, 56). Essentially the same mentality informed a protest in my own small Canadian city, Thunder Bay, where moneyed rural residents of virtual gated communities successfully stopped a major wind farm development in 2014. Property value, not climate change mitigation, was clearly their overwhelming motivation – although, of course, it was all carefully packaged in the transparently bogus terms of health and nature.

KLEIN ON NIMBYISM

[C]ommunities are simply saying "No." [...] And not just "Not in my Backyard" but, as the French anti-fracking activists say: *ni ici, ni ailleurs* – neither here, nor elsewhere. In other words: no new frontiers. Indeed the trusty slur NIMBY has completely lost its bite.
(Naomi Klein 2014, 335)

At last citizens in the West have learned to wrap personal outrage in the colour of a green planet. And so we can finally embrace Indigenous Rights and Indigenous Peoples as a convenient, effective, but entirely cynical way

"The Price of Oil" (2005), Andy Singer

of stopping unwanted developments in our own backyards. At last we can acknowledge the harms caused by colonialism. But if the oil wasn't literally flowing down American streets and beaches; if the water wasn't literally combusting along rivers and from taps in Australia and North America; if the railway cars weren't literally exploding in the downtowns of our small towns, as they did in Lac-Mégantic, Quebec; if the awful development, in short, hadn't shifted into our own backyards in places like Denton and Thunder Bay, well, then very few people indeed would be bothered enough to protest anything, let alone write or read essays and best-selling books about it.

LEAP MANIFESTO (2015)

There is *no longer an excuse for building new infrastructure projects that lock us into increased extraction decades into the future.* The new iron law of energy development must be: *if you wouldn't want it in your backyard, then it doesn't belong in anyone's backyard.*

Seventh feature of the Anthropocene condition: the hypocrisy of the rich has finally begun to overlap with the suffering of the poor. I call this unfortunate but apparently necessary development *the democracy of suffering.* More about it in section 3.

6 War on the Future

Klein echoes Frank's revisionist interpretation of the 1960s (without acknowledging it) in *This Changes Everything: Capitalism vs. the Climate* (2014), arguing that climate denial is the direct offspring of neoconservative politics as sponsored by the economic doctrine of neoliberalism. She writes:

Many of these [neo-con] institutions were created in the late 1960s and early 1970s, when US business elites feared that public opinion was turning dangerously against capitalism and toward, if not social-ism, then an aggressive Keynesianism. In response, they launched a counterrevolution, a richly funded intellectual movement that argued that greed and the limitless pursuit of profit were nothing to apolo-gize for and offered the greatest hope for human emancipation that the world had ever known. (38)

By the 1980s the lionization of greed was the dominant fashion – as represented by "Dynasty" and "Wall Street" on the small and big screens of North America. And so was "market fundamentalism" (20–6), the religion of capitalism, damn the costs. Already by the late 1980s neoconservatives had successfully spread the fatalistic belief that "there is no alternative" (TINA), Margaret Thatcher's motto for a world stripped clean, not only of the nuisance of society, or, in her case, of sociology, but of *the future*. Hence the nearly pandemic spread of US boomernomics around the world, Kant's dream of mass education and enlightenment reimagined as Coca-Cola, globalization, and American financial might.

The triumphs of boomernomics were formally elevated to philosophy in an essay of 1989 by political scientist Francis Fukuyama and then again in 1992 in his book-length extrapolation of the same thesis. According to *The End of History and the Last Man*, liberal democracy had finally reached what Hegel, as filtered through the Marxism of Alexandre Kojève, called the "end of history." For Fukuyama, liberal democracy had, with the fall of the Berlin Wall in 1989 and the collapse of the Soviet Union in 1991, final-ly unified the world under its umbrella. Liberal democracy, not Marxist dreams of utopia, was the true end of history: not communism but the freedom of the market place.

Naturally the critics of capitalism were incensed by the self-congratu-latory back-slapping that lasted throughout the 1990s and a bit beyond. By 2018, though, even Fukuyama has changed his tune – openly criticizing de-regulation, implicitly blaming neoliberalism, and explicitly praising aspects of both socialism and Marxism (see Eaton 2018). But in a way Fukuyama 1.0 was right. The fall of the Berlin Wall really was the end of history, not as the grandiose fulfillment of absolute freedom in liberal democracy but as the end of capitalism itself. For within twenty years boomernomics proved itself to be nothing less than the *reductio ad absurdum* of capitalism, its reck-less death race made abundantly clear by the relentless volatility of markets after 1989, by the dot-com bubble of the mid to late 1990s, by the financial

Checkpoint, courtesy of Pixabay

meltdown of 2008 and its interminable zombie afterlife, and by a capitalism that has very much facilitated climate catastrophe detrimental even to its own concrete existence. In short, the ends have vitiated the means.

So much, then, for the future. Instead we behave, in Gardiner's terms, as though "our [current] interests have absolute priority over the interests of the future: any interest of ours (however trivial) is sufficient to outweigh any interest of theirs (however serious)" (2011, 21). This is of course nihilism.

PEAK CAPITALISM?

The capitalist system is at present stricken with at least five worsening disorders for which no cure is at hand: declining growth, oligarchy, starvation of the public sphere, corruption and international anarchy.
(Wolfgang Streeck 2016, 72)

As the German economic sociologist Wolfgang Streeck argues, the forces of opposition may have been vanquished by capitalism (as in Soviet Russia), but these forces have always, as a rule, been the motor of change and innovation within capitalism – forces that made capitalism capitalism, that is, made it dynamic. But "capitalism without opposition," as Streeck says, "is left to its own devices, which do not include self-restraint" (2016, 65). Consequently capitalism is dying "from an overdose of itself," a victim of its own success.

We need to be blunt about this. Today a suicidal and therefore final form of capitalism is killing itself even as it kills the planet. There are no more contradictions to resolve and then exploit. As the historian Richard Smith puts it in "Green Capitalism: The God That Failed" (2011), capitalism requires a state of constant growth, while the fate of life on Earth requires massive contraction and degrowth. The "growth fetish" is in turn linked to rising carbon dioxide rates – an easy measure of how much capitalism has externalized its costs onto the planet (see Hamilton 2010, 32). Just a short time ago informed people drew a line in the sand of carbon emissions at 350 parts per million in the earth's atmosphere. In April of 2018 the measure was actually over 410 parts per mission, the highest levels in 800,000 years – and well into uncharted, highly dangerous territory for life as we know it (see Mooney 2018). It will rise higher yet.

[T]he logic of capital is predicated on infinite growth and expansion. The profit generated by private firms is perpetually reinvested into new production, which requires more land, and land, historically, was acquired through any means necessary. *This is why capitalism, colonialism, and climate change are inexorably bound up with one another: the three faces of a mutually reinforcing system of violence that is killing our planet.* This continues in the 21st century through the violation of indigenous land rights as pipelines and other carbon infrastructure are created on ancestral lands without the consent of the first peoples.
(Shiv Raveendrabose 2018, emphasis in the original)

REQUIEM FOR A SPECIES?

[T]he chances of stopping warming at 2C above pre-industrial levels are virtually zero because the chances of keeping concentrations below 450 ppm are virtually zero. In fact, in 2007 the concentration of greenhouse gases reached 463 ppm, although when the warming effect is adjusted to account for the cooling effect of aerosols the figure falls to 396 ppm. Only air pollution is protecting us. (Clive Hamilton 2010, 12)

The implications are obvious. Life requires the collapse of capitalism, since it is fundamentally incompatible with existence over the long term. This is why serious people argue that fossil fuels should be left in the ground. As the British public intellectual George Monbiot says, "There is only one form of carbon capture and storage that is scientifically proven, and which can be deployed immediately: leaving fossil fuels in the ground" (2016). That's exactly right. Fossil fuel, the lifeblood of capitalism, is what's reshaping the future as inhospitable and dystopic. Smith characterizes the dilemma perfectly: "*it is difficult to see how we can make the reductions in green house gasses the scientists tell us we have to make to avoid climate catastrophe unless we abandon capitalism*" (2016, 126; his emphasis).

CARTER'S "FIRESIDE CHAT"

We must look back into history to understand our energy problem. Twice in the last several hundred years, there has been a transition in the way people use energy.

The first was about 200 years ago, when we changed away from wood – which had provided about 90 percent of all fuel – to coal, which was much more efficient. This change became the basis of the Industrial Revolution.

The second change took place in this century, with the growing use of oil and natural gas. They were more convenient and cheaper than coal, and the supply seemed to be almost without limit. They made possible the age of automobile and airplane travel. Nearly everyone who is alive today grew up during this period, and we have never known anything different.

Because we are now running out of gas and oil, we must prepare quickly for a third change – to strict conservation and to the renewed use of coal and to permanent renewable energy sources like solar power.

The world has not prepared for the future.
(President Jimmy Carter, 18 April 1977)

Our oil-, gas-, and coal-fuelled capitalism, once a cheap and easy solution for generating global wealth, is logically unsustainable. And that's why hybrid positions, such as "green capitalism," are chimeras – pleasant fantasies. The truth is that newly built electric cars, for example, may be fine for *the economy* but they are even more harmful to the environment than maintaining used cars. Why? Because the environmental harm caused by producing cars decades ago has already been absorbed and cannot be undone now. The rampant growth of the new car industry is a cause of new and ever-increasing carbon emissions, and producing new electric car batteries requires the extensive mining (and smelting) of nickel, iron, copper, and rare earths, a devastating process that poisons soils and rivers anew. "I would not be surprised," Smith deadpans, "if the most ecological cars on the planet today are not those Toyota Priuses or even the Chevy Volts with their estimated 7–10 [year] lifespan, but those ancient Fords, Chevrolets, and Oldsmobiles cruising around the streets of Havana" (2016, 133).

In Klein's estimation, "our economic system and our planetary system are now at war" (2015, 21). And the planet is losing – because the climate deniers and liars have, until very recently, won all the big battles. Of course the United States is on the front lines of disinformation. It has been reported that "at least 180 congressional members and senators are declared climate deniers." Moreover, according to the Australian scholar David Schlosberg, "They've received more than US$82 million in campaign contributions from the fossil fuel industry and its partners" (2017). It's very clear that boomernomics has set short-term self-interest against long-term self-preservation. And every thinking person knows this is bonkers.

One philosopher of this ethical quagmire is Stephen Gardiner. He claims that a combination of factors, primarily globalization, intergenerational conflict, ecological destruction, and inadequate theoretical frameworks, has created a "perfect moral storm" – the name of his tome of 2011 (see also Gardiner and Weisbach 2016, 24–39; Shue 2018). "A central component of this perfect moral storm," he repeats in *The Washington Post*, "is the threat of a *tyranny of the contemporary*, a collective action problem in which earlier

Oil field, courtesy of Pexels

generations exploit the future by taking modest benefits for themselves now while passing on potentially catastrophic costs later" (2016). The future, rendered merely abstract, is thereby thrown away like so much plastic garbage. One is certainly right to wonder, with Byron Williston (2015), if future generations will, or should, ever forgive us.

But there's a hidden opportunity here. We could try to overcome these practical and moral problems. We could usher in both energy and moral revolutions – and thereby change human consciousness. Oxford political scientist Henry Shue (2018) puts it beautifully: "what an exciting privilege it is to belong to a set of pivotal generations with the opportunity to carry out a once-in-a-civilization transition, one of history's great revolutions, rivalling the Agricultural Revolution and the Industrial Revolution, whose dark side we are confronting." Or as Dale Jamieson (2014) puts it, today we have the opportunity to live "gracefully" and with purpose (200; cf. 8).

7 Conditions of Existence

I'm claiming that the driving force of the Anthropocene condition is a justifiably narcissistic question: the question of human existence, of the human subject, in relation to an increasingly inhospitable Earth. In what ways has the Anthropocene changed or is in the process of changing human subjectivity?

One answer was sketched in a *New York Times* article of 2013 for "The Stone," a forum for philosophical reflection about contemporary life, by former US army private (and doctoral student of English). In "Learning How to Die in the Anthropocene," Roy Scranton applies the lessons of his war experiences in Baghdad (2003–04) to better understand his American home in the wake of Hurricane Katrina. What Scranton finds is climate-induced "shock and awe" (2015, 14) and, moreover, an America incapable of managing its disasters, unprepared to address the precipitating cause in climate change, and blissfully unaware that conditions on the ground have come to resemble those in war-torn Baghdad. To register the seriousness of the problem, he cites the warnings of other military men – such as Admiral Samuel J. Locklear III, Tom Donilon, and James Clapper – and dire warnings from the World Bank.

"US Carbon Footprint" (2009), Andy Singer

Scranton explains how he managed his own existential fears in Baghdad by embracing death, essentially by thinking philosophically about mortality. On the basis of that experience, Scranton prescribes a philosophical analysis of death, not just on the individual level but also on the collective or species level, for those of us "dying in the anthropocene." "The biggest problem we face," he claims, "is a philosophical one: understanding that this civilization is *already dead*." And again, echoing his title, "If we want to learn to live in the Anthropocene, we must first learn how to die." This project he takes up again in a small book of 2015, *Learning to Die in the Anthropocene: Reflections on the End of a Civilization* – fleshed out with lots of data, witty commentary, and more thoughts about philosophy and death. It is generously blurbed by prominent authors in the field, including Klein, but here's my own:

> Despite Scranton's sentimentality about the past, flakey allusions to human "vibrations," ponderous twaddle about "photohumanism," and vast overselling of the humanities and the power of narrative, *Learning to Die* may be the only book one needs to read in this exploding field. For Scranton demonstrates that anthropocenity is less about quantitative analysis than it is about "soft" qualitative interpretation. The Anthropocene Condition is, in short, about human beings generating meaning in a perilous state of decline. Consequently it's about doing philosophy, being wise, and interrupting the bullshit. That's enough for any book to demonstrate, let alone a brief one.

THE BIG FEAR

We're scared that once you get past 2 degrees, the planet's own internal mechanisms kick in. The population comes down like a stone. A complete collapse. You lose the civilization entirely. (Astrophysicist Adam Frank, cited in Hedges 2018)

PENTAGON ON CLIMATE

Even if sudden shifts in the climate do not materialize, gradual shifts in climate could nonetheless spark surprising secondary effects – such as a massive release of gases from melting permafrost, persistent megadroughts, extreme shifts in critical ecosystems, emerging reservoirs of new pathogens, or the sudden breakup of immense ice sheets. The national security implications of such changes could be severe. (Pentagon's "Implications for US National Security of Anticipated Climate Change" 2016, 11)

Actually, though, Scranton's recourse to existentialism already has a pedigree in the climate literature. Klein, for example, addresses our existential situation throughout *This Changes Everything*, often very plainly. The accretion of traumatic climate events, she says, signifies that "climate change has become an existential crisis for the human species" (2015, 15). Or as the Australian professor Clive Hamilton puts it in a book of 2010, "Humanity's determination to transform the planet for its own material benefit is now backfiring on us in the most spectacular way, so that the climate crisis is for the human species now an existential one" (2010, xiii). As such we must allow ourselves to "enter a phase of desolation and hopelessness, in short, to grieve" (211). More recently people have begun to speak not just of ecological trauma but also of "ecological grief."

And then there's American journalist Elizabeth Kolbert's Pulitzer Prize–winning book, *The Sixth Extinction: An Unnatural History* (2014), which argues that the Anthropocene names a planetary die-off as last seen during the extinction of dinosaurs about 65 million years ago. Like some other books in this now burgeoning field, it's based on a previously published essay – this one appearing in *The New Yorker* in May 2009 but punctuated with a (purely rhetorical) question mark, "The Sixth Extinction?" In some ways Kolbert's diagnosis is more far-reaching than analyses of capitalism, oil extraction and consumption, coal use, natural gas fracking, and rising methane levels (especially with the melting permafrost in the far north), and so is even more pessimistic than the others. Kolbert describes the systematic extinction of ancient creatures, like frogs and bats, directly correlated to the expansion of human beings into natural habitats.

For example, Kolbert explains how nomadic human beings spread pathogens into new ecosystems; introduce new species through the discharge of ballast water on transport ships; introduce predators, like rats, to unsuspecting populations; overfish and overhunt once-rich reserves; dam rivers and deforest large areas for agriculture, industry, and habitation; and, of course, introduce pollutants at every stage in the production of consumer foods and goods. Some species simply cannot survive contact with human beings – and it often has nothing to do with carbon emissions and the acidification of our lakes and oceans. The world population has almost doubled since 1970, and the planet is having trouble absorbing the activity of more than 7 billion people. The feedback effect is the death of biodiversity all over. Hence the dramatic subhead of Kolbert's original essay: "There have been five great die-offs in history. This time, the cataclysm is us."

Journalist Paul Masson is just as broad-minded in *Postcapitalism: A Guide to Our Future* (2015), carefully piling on factors that contribute to our current mess. A big one is demographic trends, which he calls a "timebomb." In the West, aging populations and tanking birth rates have coincided with a capitalism in free fall. One result is that pension funds are at serious risk, which is why they have been slashed so aggressively throughout the Western world over the last decade. Getting old won't be any easier in the coming decades. But at least the countries most responsible for carbon emissions – and by a factor of between 50 and 130 times more than in the developing world (Hamilton 2010, 43) – are shrinking. Even so, the world population is projected to produce another 2.6 billion people by 2050, a disproportionate percentage in sub-Saharan Africa. For example, by 2050 Niger's population will rise from 18 to 69 million. "To find jobs," Masson writes, people "will migrate to the cities; the land, as we've seen, is already under stress from climate change. In the cities, many will join the world's slum-dwelling population, which already stands at a billion – and increasing numbers will attempt illegal migration to the rich world" (257). The overall effect, to put it mildly, will be a significant challenge to democracy (258) and ecology. And to all human life.

Laughing cops, courtesy of Pixabay

Asian cityscape, courtesy of Pexels

PENTAGON REPORT

When climate-related effects overwhelm a state's capacity to respond or recover, its authority can be so undermined as to lead to large-scale political instability. Countries with weak political institutions, poor economic conditions, or where other risk factors for political strife are already present will be the most vulnerable to climate-linked instability. In the most dramatic cases, state authority may collapse partially or entirely. (2016, 6)

The last time it took an asteroid to cause a mass extinction. This time incessant human activity, much of it organized around capitalism, will be enough. A suitable motto for the Anthropocene? Kolbert: "The Anthropocene, this time the cataclysm is us." Scranton: "The Anthropocene, this time 'the enemy is ourselves'" (2015, 85). Or, negotiating the difference: "The Anthropocene, this time we don't get out alive." It seems clear that the anthropocenic subject is a *mortified* subject.

Our preoccupation with death and mass extinction brings us to the eighth and perhaps most obvious feature of the Anthropocene condition: the return of existential angst, only now on a planetary scale.

What's especially troubling is the perverse taste we have all developed, at once masochistic and narcissistic, for this unfolding spectacle of death and destruction. Everyone has a front row seat in the Anthropocene, which looks more and more like an exhibition by the photographer Edward Burtynsky. Everyone is settling in, getting comfortable, to enjoy around-the-clock 24/7 edutainment in industrial porn, ruin porn, weather porn, refugee porn, riot porn, protest porn, disaster porn, collapse porn, recovery porn, financial porn, election porn, impeachment porn, etc. In fact there is no doubt about it at all: thanks to modern communications and an insatiable desire for spectacle, the future promises to be excessive, to wit, pornographic. Or rather, to be more exact, *masturbatory*.

And philosophy porn? Yes, that too.

"Industrial Porn" (2010), Todd Dufresne

8 Strategic Opportunism

The clever cheekiness of Klein's argument is that she openly assesses the Anthropocene condition in terms previously reserved for the abusive "shock doctrines" of capitalism. In her previous bestseller, *The Shock Doctrine* (2007), Klein demonstrates how neoconservatives utilized economic, social, political, and environmental crises to push through neoliberal reforms and policies, ultimately in the service of creating "disaster capitalism," the name Klein gives to the cynical profiteering achieved by privatizing "the commons" while coarsening and militarizing society. The link between the shock and climate books is precisely this idea that a shock is a terrible thing to waste, a cynical view championed by opportunists like Edward Greenspan but echoed by Klein as a rallying call for a new environmental movement. In Klein's words, the shock of climate change could perhaps incite a "People's Shock, a blow from below" (2014, 10). The existential threat of climate change is thus a wake-up call for a new society built not by the political class but by the everyday victims of capitalism. By the 99 per cent. It's a reasonable, hopeful idea that I echo below.

Of course Klein's own opportunism concerning the shock of climate change renders her analysis just as motivated as the neoliberal (or, if you prefer, neoconservative) opportunism she decries, only now on the side of aggressive Keynesianism, socialism, or communism – she isn't clear which. She thus represents everything that neoconservatives fear, namely, an unapologetic return to the eclipsed ideals of the 1960s. At the same time, not every case of opportunism is equal. For if climate change is real, then Klein's *cri de coeur* is merely correct. Moral outrage and calls for change are obviously very much needed – notwithstanding the lingering disinclinations of postmodernity. Consequently, the most important parts of *This Changes Everything* are not really the rehearsal of facts and figures about climate change, the debate about Green Economics and its detractors, and so on. The most important parts of the book are her exploration of concrete wins "from below," typified by "Cowboy and Indian" alliances to stop development projects (often oil pipelines) and her closing remarks about the possibility of a fundamental change to human consciousness, what I'm calling the birth of the anthropocenic subject. The first part is practical, the second practico-intellectual (or spiritual). As Klein puts it, the task of delegitimizing the capitalist worldview is all about changing patterns of thought (460–1). That powerful lesson reverberates across the literature.

Oil rig explosion, courtesy of Pixabay

And so Klein argues that "We need a Marshall Plan for the earth" (5) – a massive and well-organized response, as in the wake of the Second World War, to the problem of global warming. That's a good idea. But the vested ideological interests of disaster capitalists make this plan absolutely pie-in-the-sky idealism. True, the "existential crisis that is climate change" could activate collectivist values long suppressed in the West, and it could, as she says, "provide us with a chance for a mass jail break from the house that their ideology built" (63). But the timeline required for such massive change is not feasible, at least for anyone wanting to preserve as much civilization as possible. Wholesale economic change needs to be massive and immediate. And Klein hangs on to this possibility: "culture can shift quickly" (60). But it won't – at least not without a push. More about the "mass jail break" from capitalism, and the push needed to get us there, in section 3.

9 Of Human Nature

In a review article of 2015, Rebecca Tuhus-Dubrow perceptively characterizes a major debate at the heart of literature on the Anthropocene. "Is it humanity per se," she asks, "that has brought about these massive disruptions, or is it a very specific economic and political system, benefitting a very small subset of people, that is responsible?" Kolbert is mostly in the former camp (the word "capitalism" doesn't even appear in her index), while Klein is mostly in the latter camp. But Klein is also careful to indict claims about our supposedly Hobbesian nature and to that extent broadens her analysis to include grander problems about the fundamental characteristics of human existence. This is sensible since, arguably, the conservative belief in immutable human nature is the most intractable problem of all. For it is an ideology that naturalizes human destructiveness and greed, and, to that extent, naturalizes capitalism. "We will need to start believing," Klein pleads, "that humanity is not hopelessly selfish and greedy – the image sold to us by everything from reality shows to neoclassical economics" (2014, 461). Or as she declares as the title of an article of 2018, "Capitalism Killed Our Climate Momentum, Not 'Human Nature.'" Simply put, we must start to believe again in a *collective* and *collaborative* future, one that transcends the old political options. Klein spells it out for us in the 2018 article: "a new form of democratic eco-socialism, with the humility to learn from Indigenous teachings about the duties to future generations and the interconnection of all of life, appears to be humanity's best shot at collective survival."

Of course the debate between nature and nurture is ancient. Minimally it accounts for pessimist and optimist beliefs about the possibility of change. Freud is a classic (albeit much misunderstood) case of a pessimist heavily invested in the idea that we cannot really alter our given human natures. Hence his famous distrust of the *furor therapeuticus* and characteristic irony about the goal of psychoanalytic therapy: to transform "hysterical misery into common unhappiness" (1895, 305). This is also why neo-Freudians, after Freud's death, spent so much energy trying to alter his basic teaching, stripping it of its marriage of biologism and late Romanticism. Unsurprisingly, therapists want treatment to make people happy and, if possible, to cure their neuroses; they want to help create, if only on the individual level, a new future. Some Freudians therefore turned to Marx. For unlike Freud's pessimism about change, Marx argues that if you change the economic structure you will automatically change human nature.

For Marx, change-revolution-progress are all predicated on the essential mutability of human nature. This is why progressives, often in the spirit of Marx, favour laws and bylaws that manipulate or manufacture human nature. Progressives, for example, may advocate for restricting the number of liquor licences granted within a given business district. For when you cut access to taverns or liquor stores you see measurable drops in crime and various forms of civil misconduct. And this, by the same token, is why conservatives hate progressives (twentieth-century "liberals") so much: it all smacks of a nanny-state paternalism that dares legislate what is better left to individual free choice or, less politely, to Darwinian-styled survival of the fittest. Stated otherwise, progressives want to help improve people, while conservatives want to leave people to rise or fall on their own (given, innate) merits.

Klein is a progressive, and in truth the best thinkers of the Anthropocene condition are the same. Why is that? Because only the progressive viewpoint is optimistic about change and on that basis is willing to challenge the dangerous status quo. Only the progressive position – even when it's condescending, wrong, and laughably righteous – aims to help people, aims to open up the future. Conservatives want people to help themselves and, when they fail in this task, suffer the justified fate of their failure. Hence the shocking mean-spiritedness of Ayn Rand and her ilk – often called libertarian egoists and "Objectivists," but more simply and less mysteriously called *assholes*.

"Dollar Sign" (2017), Dwayne Booth

10 Fatalism: Marketing Strategy & Psychology

Fatalism about climate change is therefore highly problematic for progressives. For it means that nothing can be done to prevent or mitigate a catastrophic future. It implies (small "c") conservatism. One form of fatalism says that the extreme weather events we're experiencing are natural, cyclical, and so unavoidable. That's the position held by enlightened deniers, who at least concede that something very bad is happening. The other form of fatalism, a kind of apocalypticism, is just as dangerous and is actually more common among progressives: human-generated climate change, while obviously real, is so far advanced that it's too late to mitigate and/or avoid. As Scranton puts it, "We're fucked. The only questions are how soon and how badly" (2015, 16).

WE'RE FUCKED

They held this conference in Oxford and I went along. As the conference started, there was a kind of suppressed emotional intensity, except in the coffee breaks. It was then that I would buttonhole a couple of scientists and say: "Well, you know we're speculating about this. But what do you really think is the situation?" And one of them just looked at me and said: "We're fucked." (Clive Hamilton, in Kennedy 2017)

A pure form of fatalism is famously expressed by the freethinking British scientist James Lovelock. Lovelock is best known for the "Gaia hypothesis," first advanced in 1979, according to which the Earth is a living, self-regulating, self-correcting, single system that, for any given human input, responds with feedback. At first ridiculed as an old fashioned return of animism (see Doolittle 1981, 58–62), and always criticized for being merely metaphorical (Gould 2007), Gaia is now widely considered a starting point for the earth sciences (see Hamilton 2010, 149–51). It also inspired further musings and books from Lovelock, including prophetic declarations about the future. In support of a new book of 2006, *The Revenge of*

Gaia, Lovelock wrote an op-ed for *The Independent* entitled "The Earth Is About to Catch a Morbid Fever That May Last as Long as 100,000 Years." A self-described "planetary physician," Lovelock warns readers "that you and especially civilization are in grave danger." "We have given Gaia a fever," he writes, "and soon her condition will worsen to a state like a coma." And again, "before this century is over billions of us will die and the few breeding pairs of people that survive will be in the Arctic where the climate remains tolerable."

In an interview at *The Guardian* two years later Lovelock fully embraced his role as modern Cassandra, declaring that we had passed the tipping point when severe global warming is not just inevitable but irreversible. As such, green efforts like recycling, ethical consumption, carbon offsets, and renewable energy schemes were pathetic and delusional wastes of time. His frank advice to his interviewer: "Enjoy life while you can. Because if you're lucky it's going to be twenty years before it hits the fan."

Just four years later, however, the ninety-two-year-old Lovelock, ready to promote another Gaia book, retracted his previous prediction as "alarmist." Global temperatures hadn't risen as high or as fast as expected. That said, the outlook was still very bad – it was just that Lovelock's original disaster timeline was wrong. Most telling of all is Lovelock's droll remark that he should have been more cautious in *Revenge of Gaia*, "but then that would have spoilt the book." It wouldn't have sold as many copies, either.

Arguably Lovelock's droll tone rises to the level of a ninth feature of the Anthropocene condition: the irruption of gallows humour and giddy amusement, inevitable counterparts to existential angst about the future.

REBUTTING LOVELOCK

I have to say that I agree with some of his targets for criticism – carbon offsetting, for example – but I don't agree that we shouldn't make an effort to analyse the varying impacts of our lifestyles and at least try, even if we are sometimes destined to fail or later find it caused an unwanted side-effect, to change direction when we know our lifestyles can cause ripples of negative influence around the world – be it climate change, degraded habitats or social injustice. This is the mindset that underpins ethical living, eco living, sustainability – call it what you like – despite whatever the knockers may say.

But what I find more unpalatable about the tone of Lovelock's comments is his barely disguised glee that we are going to get what we deserve for not listening to his warnings about our bespoiling of the atmosphere – an 80% reduction in global population levels by the end of the century. There is more than an air of the Old Testament about what he says, namely, that we are going to be punished without mercy for our sinful ways. He may well be right, but why the "told you so" tone? (Leo Hickman, 2008)

11 Fatal Attraction

Reporters often remark about a strange disconnect between Lovelock's personal good humour and his fatalism. But Lovelock insists he's an optimist, and in 2008 tells us why: because the spectre of human extinction is invigorating. His reasoning, coherent yet astonishingly infantile, comes from his own experience of the hardships of the Second World War. From *The Guardian* in March 2008:

> Humanity is in a period exactly like 1938–39, he explains, when "we all knew something terrible was going to happen, but didn't know what to do about it." But once the second world war was under way, "everyone got excited, they loved the things they could do, it was one long holiday … so when I think of the impending crisis now, I think in those terms. A sense of purpose – that's what people want." (Aitkenhead 2008)

Three stacks, courtesy of Pexels

Of course it's true. Everyone wants a sense of purpose in their lives. That's why we turn to ancient investments like religion, mysticism, art, philosophy, drugs, sex, and affairs of the heart, and to modern investments like science, music, video games, fashion, yoga, artisanal beer, work, vintage cars, social media, conspicuous consumption, and so on. It's also true that nothing sharpens the mind, or cracks open the soul, like the spectre of death. But whether or not the spectre of death also counts as "one long holiday" is surely another matter.

FREUD ON THE USES OF WAR

Life is impoverished, it loses in interest, when the highest stakes in the game of living, life itself, may not be risked. It becomes as shallow and empty as, let us say, an American flirtation, as contrasted with a Continental love-affair in which both partners must constantly bear its serious consequences in mind … [W]ar is bound to sweep away this conventional treatment of death … Life has, indeed, become interesting again. (1915, 290–1)

The claim that suffering, death, and imminent extinction are a heightening of life is obviously cocky, macho, and celebratory. But it's not the only lesson one can draw from catastrophe. Take Schopenhauer's droll response to the widespread miseries of post-Napoleonic Europe. In "The Sufferings of the World" he contends that ubiquitous everyday suffering should be enough to make us more indulgent of "fellow sufferers." More substantially he argues that suffering, like a ship's ballast, is precisely what keeps people true, sane, grounded, real. In short, suffering has its uses. In contrast to Lovelock, Schopenhauer's view is passive and depressive, a step toward what would eventually become existentialism.

I thank God I am not young in so thoroughly finished a time. (Goethe 1824)

Tenth feature of the Anthropocene condition: individual and group psychologies in our time flit between mania and depression as exemplified in the extreme examples of Lovelock and Schopenhauer. As for a middle point of exalted depression? That might be represented by Franco Berardi, one of the most pessimistic thinkers of the future working today. More about him later.

12 Love & War

Lovelock isn't the only one to invoke war metaphors to grasp or measure our current predicament. In his "Telling Friends and Foes in the Time of the Anthropocene," French social philosopher Bruno Latour deploys and explicitly defends the use of war rhetoric – and along with it the necessity of "mobilization." The immensity of the problem, he argues, requires an immense political *and* scientific reaction. It is in a similar spirit that Naomi Klein invokes the Marshall Plan as a model for imagining what kind of action is needed to combat global warming. More recently yet, David Wallace-Wells has insisted that "the World War II mobilization metaphor is not hyperbole" (2018) – a claim he repeats again in *The Uninhabitable Earth* (2019, 20, 23).

War rhetoric is invigorating and telegraphs the calamity of existence in the Anthropocene. The idea is to enlist as many citizens to a global cause and at the same time remind us that such mobilization has been accomplished in extraordinary circumstances in the recent past. But this time we are talking about a "war" for the love of existence, for the love of Earth, for the love of future generations – an oxymoronic mash-up of war and love that adequately captures the ironies endemic to our time.

Arguably this war rhetoric is a stark measure of our desperation. After decades of "war" on drugs and then "war" on terror, both interminable because impossible to win, a new "war" on climate change courts predictable failure from the outset. For although war rhetoric is titillating, it can't possibly achieve the stated objective: mobilization to effect deep, fundamental change on a global scale. The deployment of war rhetoric also erodes the meaning of (let's just say) nonmetaphorical war, the kind where people kill and are killed at the hands of other people.

Yes, of course we know it's true: capitalism and climate change are the real structural killers, the generators of poverty and fatal inequalities, pension

Forest fire, courtesy of Pixabay

shortfalls, calamitous impoverishments everywhere, homelessness and statelessness, rampant disease, climate-caused catastrophes, extinctions, and also wars. A "war" on people and planet has been waged practically since the inception of capitalism, notwithstanding the significant number of relatively free and affluent winners from capitalism. We can now see that it has been, from a global perspective, a losing war. But the demise of capitalism in our time has almost nothing to do with these critiques, almost nothing to do, for example, with Marxist resistance. The truth is that capitalism never lost the war with anticapitalists; it won, and won spectacularly. As Streeck says, the success of capitalism has been its undoing. Worst of all, there's nothing waiting in the wings to take its place – save for some form of harsh fascism or, less likely, benevolent socialism.

> Nothing we need to do to stop climate catastrophe is politically realistic. But what's truly unrealistic is thinking that the future is just going to be brighter on its own, that future generations will solve their own problems. There's no time left for … incrementalism. The future – our future – is at stake. (Alyssa Battiston 2018)

Perhaps the real problem with a "war" on climate change is that it mistakes Earth, mistakes Gaia, for a battlefield. But it's not a battlefield and even less so an enemy (leaving aside, for the moment, the beliefs of transhumanists who hope to "win" by transforming human biology). This is where Scranton's depressing audit of anthropocenity is helpful: the problem today, recall, is that we are the problem. We are the ones, as 350.org says, who must be convinced to leave 80 per cent of all oil reserves in the ground if we are to stay below the two-degree threshold of climate disaster. And even that threshold is optimistic. If so, then war rhetoric has to have a quite different target: a genuine "war" on climate change must be a "war" on human beings or, more precisely, a "war" on the way human beings have fashioned themselves over the last three hundred years. A war on human nature.

AN ASTERISK ON OPTIMISM

*Even if we do manage to keep 80% of fossil fuels
in the ground, a world that's 2°C warmer is going to
be a much different, scarier place. We're only at +1°C
now, and we're already seeing more storms, flooding,
heatwaves, drought, and island nations at risk of going
underwater. +2°C is going to mean a lot of human
suffering, and tremendous damage to the planet.
(350.org, qualifying the "Good News")

And so we turn a page, once again, on Plato and ancient Greece. In Book 2 of the *Republic*, Plato introduces soldiers to defend the decadent way of life in the "city in high fever," a.k.a. Athens. This is what his interlocutors, Glaucon and Adiemantus, actually desire: more of the same, more cake and ouzo, more Athenian decadence. But Plato's soldiers, given an education in wisdom loving, are given a new mandate by Plato. These new and improved soldier-philosophers are actually redeployed to defend the just city from its less than just citizens – to defend humanity from its own base desires. In short, soldiers introduced to guarantee the decadent way of life actually go to "war" against the decadence of the city's own citizens. Hence the greatest, most difficult law of the Republic: philosophers must be king or queen. As for the unjust citizens of Athens? Unable to discipline themselves, philosophers and soldiers will discipline them instead. This is philosopher as disciplinarian and moralist.

I doubt this is the future we want. Like Glaucon, we still want "the city in high fever," Athens or New York or Tokyo. More ouzo – and more avocadoes on toast. Yet is it possible that this is the future we need?

The moment we fully accept the fact that we live on a
Spaceship Earth, the task that urgently imposes itself is
that of civilizing civilizations themselves, of imposing
universal solidarity and cooperation among all human
communities … (Slavoj Žižek 2017)

"Chocolate World" (2018), Todd Dufresne

13 Ball of Confusion

There is a lively debate about what should count as the beginning of the Anthropocene. Let's take the main origin claims in reverse chronological order. Some point to "the great acceleration," ca 1950–present (Steffen et al 2007); the Industrial Revolution, ca 1760–1840; the convergence of old and new worlds associated with the "Orbis spike" of 1610 (Lewis and Maslin 2015); deforestation and mining practices about 2,000 years ago (Certini and Scalenghe 2011); a methane spike related to rice farming in China about 5,000 years ago (Purdy 2015, 2); agriculture and deforestation practices about 8,000 years ago (Ruddiman 2003); and domestication and agriculture practices about 10,000 years ago (Smith and Zeder 2013). Some authors even point to the origin of the human conquest of fire as the first human-made carbon pollution and thus peg anthropocenic climate change to a near-determinist trajectory that begins in prehistorical times.

Arguably these gambits, issuing from the scientific literature, miss a critically important feature of the Anthropocene, namely *anthropocenity*, which is not the moment clueless human beings first generated carbon pollution or, by cultivating the land, caused methane levels to rise. Anthropocenity is not really about the literal origins of this epochal shift – any more than it's about the fact of anthropocenic climate change. It's about human consciousness *of* this shift, *of* this fact. Once again, it's about the subjective recognition of an entirely new *condition*. It's about human beings. Or, in Chakrabarty's (2015) terms, it's about a "homocentric" and "moral" perspective in the face of the "zoecentric" and "causal" one. In short, our eleventh feature: the Anthropocene condition points to a revolution in *thinking*, to a shift in human *consciousness*, one that comes long after the Anthropocene proper was born.

The great acceleration is now the most widely accepted birth date for the Anthropocene proper. But the *condition* of living in the Anthropocene, anthropocenity, also has a discernable birth date: the period of time between 24 December 1968 and 7 December 1972. On the former date, Apollo 8 astronaut William Anders took the photograph *Earthrise*, an image of the Earth rising over the partially visible surface of the moon. On the latter date, Apollo 17 astronauts took the (jointly credited) photograph *Blue Marble*, Earth seen as a complete sphere set against the void of space. Each image is among the most widely circulated photographs in history, in part because NASA left them in the public domain (and so free to reproduce) and in part because they were (and in some ways remain) cultural touchstones for what it means to be a human being in the contemporary world.

Cultural geographer Denis Cosgrove has written the definitive analysis of the "contested" meanings of these two photographs in the Western context, which he considers to be markers of newly emerging worldviews. According to Cosgrove, the photographs haven't generated adequate analysis because of their ubiquity (1994, 276) and because of our unthinking belief in the neutral objectivity of photographs and their ideal eyewitness, the astronaut (278). What goes unchallenged, he claims, is how the photographs reinforce *and* subvert Western discourses of power, expansion, imperialism, mastery, control, masculinity, and the gaze.

These competing discourses exemplify what Cosgrove calls the "One-World" and "Whole-Earth" concepts. The One-World concept is epitomized by the influential "Riders on the Earth," a short article written by the American poet Archibald MacLeish. Originally published on Christmas day in 1968 as "A Reflection: Riders on Earth Together, Brothers in Eternal Cold," the article accompanied *Earthrise*. "To see the earth as it truly is," MacLeish writes, "small and blue and beautiful in that eternal silence where it floats, is to see ourselves as riders on the earth together, brothers on that bright loveliness in the eternal cold – brothers who know now they are truly brothers." (In later editions it was reprinted with the better known shortened title and minor but interesting changes.)

Cosgrove aligns *Earthrise* with a triumph of humanist ideals set against a Cold War posture of imperialism, only now reimagined as American globalization and financial might. In this vein he highlights a remarkable fact: although preceded by decades of criticism, the word "globalization" was only coined *after* the Apollo missions. Henceforth earth became "Earth," a planet or globe. In short, it's only after *Earthrise* that the concept of global "brotherhood" and "One-World" makes sense.

Conceptual push-back was simultaneously registered by the Whole-Earth concept, a blunt rejection of the rank triumphalism associated with the self-satisfied humanism of One-World. Instead of "Riders on the Earth" of 1968 we have, say, Jim Morrison and the Doors' "Riders on the Storm" of 1970: "Into this world we're thrown." Instead of appeals to human rationality we have the Frankfurt School and Heidegger-inspired rejections of instrumental reason and technology. And instead of optimism about the future based on American can-do rationality, we have newfound

Earthrise, courtesy of Pixabay

pessimism about population growth (Malthus), the recognition of myriad forms of inequality (civil rights), geopolitical overreach (Vietnam), and environmental destruction (Rachel Carson). In short, humanism in the late 1960s is countered by antihumanism. According to this darker perspective, the Apollo missions are the apotheosis of a tradition that turns subjects into objects, that reduces sacred Being to profane metrics. This is the gist of Heidegger's complaint, already in his 1966 *Der Spiegel* response to the images of Earth produced by the satellite Sputnik: the earth beneath our feet had been supplanted, at last, by an abstraction, by an Earth or globe.

14 Après nous, le déluge ...

Intellectual historian Benjamin Lazier provides a sophisticated update to Cosgrove's musings about the "Post Earthrise era," reminding us that intellectuals like Heidegger and Hannah Arendt believed that photographs of Earth from space debase human existence.

Heidegger's views are well-known. In an address of 30 October 1955, Heidegger characterized the "growing thoughtlessness," *"flight from thinking,"* and "loss of rootedness" as regrettable conditions of the atomic age (1955, 45). Stuck in big cities "chained to radio and television," enamoured by "the movies" and "picture magazines" (48), mere "calculative thinking" had eclipsed the humane value of "meditative thinking" (46). The world thus became an object, not a home. "Nature becomes a gigantic gasoline station," Heidegger complains, "an energy source for modern technology and industry" (50). Such unreflective thinking "rules the whole earth." And then he adds, "Indeed, already man is beginning to advance beyond the earth into outer space." Heidegger was dogged about this idea. Upon seeing the early Sputnik photographs, Heidegger confessed that he felt "scared."

As for Arendt, a former student of Heidegger's, she writes in a letter to Karl Jaspers, "What do you think of our two new moons? And what would the moon likely think? If I were the moon, I would take offense" (in Oliver 2015, 27). Philosopher Kelly Oliver reminds us that the first words of Arendt's *The Human Condition* (1958) "refer to Sputnik as the most important, and the most dangerous, event in human history" (27).

Lazier traces the fear of, and contempt for, these new representations of Earth to an unpublished essay by Edmund Husserl, the father of phenomenology and Heidegger's teacher. In his 1934 "Foundational Investigations of the Phenomenological Origin of the Spatiality of Nature," Husserl argues that human beings experience the everyday world beneath their feet in pre-Copernican terms. For although science long ago displaced human beings from the centre of the universe, it did nothing to unseat the phenomenological experience that the sun still revolves around us. "The fully constituted world," writes Husserl, "has its subjective departure-point and ultimate anchorage in the Ego who does the demonstrating" (224). And again, "everything comes to this: we must not forget the pregivenness and constitution belonging to the apodictic Ego or to me, to us, as the source of all actual and possible sense of being …" (230). Or as he more simply puts it on the envelope storing his essay, "*Overthrow of the Copernican theory* in the usual interpretation of a world view. The original ark, earth, does not move" (Husserl 1981, 231n1). The Apollo photographs in particular, which reduce Earth to an object, dramatically sever that remaining belief, revealing the subjective experience of phenomenal reality as the illusion it always was. With the Sputnik and Apollo pictures, the conceptual (first advanced by Copernicus) finally becomes experiential: one can indeed *see* the images of the whole Earth. No longer can one say, with Husserl, that the Earth is "certainly not perceivable in its wholeness all at once and by one person" (222).

Of course we don't really see the whole Earth at all. As Oliver says, we always only see a partial representation of that Earth, a two-dimensional photograph of one side of the Earth (2015, 22–3). It is in this vein that Heidegger complains that technology has come to so thoroughly mediate human experience that even the earth under our feet is mediated, for example, represented by means of a camera, sent over the airwaves, and finally rendered as pixels on a television screen. In the *Der Spiegel* interview of 1966, Heidegger summarizes his fear of instrumental reason and "planetary technology" at some length:

> Everything functions. That is exactly what is uncanny. Everything functions and the functioning drives us further and further to more functioning, and technology tears people away and uproots them from earth more and more. I don't know if you were scared; I was certainly scared when I recently saw the photographs of the earth taken from the moon [Sputnik]. We don't need an atom bomb at all; the uprooting of human beings is already taking place. We only have purely technological conditions left. It is no longer an earth on which human beings live today.

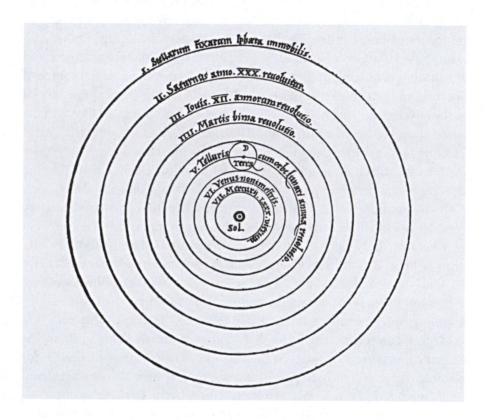

Copernicus, courtesy of Pixabay

Or as he already says in his address of 1955, the ever-changing forces of technology "claim, enchain, drag along, press and impose upon man" a stupefying "indifference toward meditative thinking" – which is, in sum, the "essential nature" of human beings (1955, 51, 56). Despite the role Heidegger played within the Nazi bureaucracy (as university rector) between 1933 and 1934, he concedes that technological rationality made monstrosities like the concentration camp possible. (These rare remarks once seemed enough to excuse his politics as merely naïve.) Like others of his generation and after, the greatest icon of technology in the mid-twentieth century was therefore the mushroom cloud – nuclear annihilation by unchecked, ungrounded, inhuman instrumental rationality. By can-do American rationality. But the slippery slope includes any technologies that alienate humanity from Being, from existence, from our *Dasein*, our being-in-the-world. Or again, functionality trumps what Heidegger calls *Bodenständigkeit*, our rootedness to earth – a sometimes ponderous feature of his worldview parodied in Hal Ashby's great movie of 1979, *Being There*, starring Peter Sellers as the dunce-philosopher.

Heidegger's critique is powerful, coherent, and in many ways convincing. But Heidegger, like the Frankfurt critics before him, doesn't get the last word on the meaning of a world mediated by technology. For Heidegger's phobic rejection of technology, like that of Jacques Ellul and others of the era, is undone by its naïve, utterly romantic view of authenticity and nature. As such it seems very much of-its-time. In fact, it was impossibly dated already in the 1960s. The truth is that many intellectuals in Heidegger's own time thought quite differently about the Apollo photographs. Lazier notes, for example, that the German historian Hans Blumenberg considered *Earthrise* and *Blue Marble* to be moments, not of alienation from earth but of self-conscious reconciliation of our finite place against the backdrop of infinite space. And there is really no doubt at all. The pictures mark the emergence of a new recognition, not of individual subjects set apart from one another, but of our existentially fraught collectivity as members of one planet – what Cosgrove calls the Whole-Earth concept. The Apollo pictures punctuate, in turn, the shift to an ecological consciousness, as registered by the growth, for instance, of the *Whole Earth Catalog* and by environmentalism more generally. It also made possible many of the hallmarks we associate with that other emergent and totalizing concept of our own time: the Anthropocene.

In the previous century Nietzsche prophetically declared the death of God, announcing the absolute and also paralyzing freedom of an existence beyond the old categories of good and evil. Human beings are the sole measure of all things. Human beings *make* value and so are impelled, in Nietzsche's terms, to revalue all values hitherto. In a way, *Earthrise* and *Blue Marble* make tangible this glimpse at absolute freedom, consolidating in two photographs the recognition that human beings are truly alone, adrift in space, one living planet among all the apparent nothingness around us. This recognition frightened people like Heidegger, who failed to see themselves in this new abstraction, but it perfectly encapsulates the dominant trends of the previous decades with a new concept – one that's neither wholly bad nor wholly good. And that is "globalization."

Let's spell this out again, very plainly, because it's important. The recognition of the "Anthropocene condition" has done for the *geological* imagination what the "death of God" did for the *philosophical* imagination and what the Apollo pictures did for the *geographical* imagination. In this sense *Earthrise* released the freedom Nietzsche imagined for the future – when it became the freedom of unchecked capitalist destruction. In a way, Heidegger's fright was realized when modern society so easily crushed the earth under our feet. But that wasn't the end of the story. The recent shift to the Anthropocene completes and also radicalizes the shift announced by the Apollo pictures, giving birth to a new concept that more than matches the frightening freedom first glimpsed by Nietzsche in the 1880s: namely, a frightening and absolute form of *responsibility* that is at the heart of anthropocenity.

Twelfth feature of the Anthropocene condition: after a century of absolute freedom to wage war and facilitate market rapaciousness for the benefit of individuals – manifest as a kind of *absolute irresponsibility* – a new kind of *absolute responsibility* has become a conceptual inevitability for human beings. After capitalism, a deluge of guilt and burdensome responsibility.

15. Answering the Question: What Is Posthumanism?

Humanism has collapsed under the weight of its own contradictions, and today the fiction of human exceptionalism has been debunked in every quarter. It therefore seems obvious that we live in an age *after* Enlightenment humanism and that the term *posthumanism* perfectly names this condition

– at least better than that other term, "postmodernism," which, as we've seen, is so vague as to be useless and, worse yet, so stripped of politics as to be dangerous.

Yet I think the term posthumanism is really a kind of branding exercise and, as such, is not just superfluous but needlessly confusing. Superfluous, first, because everything interesting that can be attributed to posthumanism can be attributed to the antihumanism advanced between the 1960s and 1990s. More precisely, it can be attributed to that French poststructuralist named Jacques Derrida. That's why the English professor Cary Wolfe's answer to the question, *What Is Posthumanism?*, is effectively *deconstruction warmed over*. And so although Wolfe's championing of "the disciplinary, institutional, ethical, and political stakes" of posthumanist thought, described as "mutational, viral, or parasitic" (2010, xix), is convincing and correct, it's just a new way of characterizing Derrida's work. To wit, it's an old way. Consequently, the term posthumanism may be good for generating new lectures, essays, dissertations, and books and may be good for bolstering claims for tenure, research grants, and promotion but it's actually the name for the future as imagined by the recent past. It's so very 1982.

More troublesome still, the term posthumanism muddies the conceptual waters. For here we have a term that also includes the hallmarks of old-fashioned humanism. Wolfe understands this problem perfectly. The popular strain of posthumanist techno-utopianism called "transhumanism" represents philosophy's oldest dream of transcendence: in the future we cyborgs will finally download our consciousness onto the web and live forever as gods of 1s and 0s. As Wolfe rightfully concludes, "my sense of posthumanism [which he calls 'genuine posthumanism' (xix)] is the *opposite* of transhumanism" (xv). And that makes good sense. Nonetheless Wolfe doesn't draw the logical inference: If "genuine posthumanism" doesn't mean anything but deconstruction and its well-known version of antihumanism, and is in fact routinely misunderstood to be an aspect of neohumanism or even super-humanism as exemplified by transhumanist ideals, then why bother championing the term in the first place?

However there's another, arguably better but more complicated, reason to jettison the term "posthumanism." Wolfe's book is predicated on the idea that "We must take yet another step, another post-, and realize that the nature of thought itself must change if it is to be posthuman" (xvi). Ok, that's interesting. But the truth is, our era is *not* really beyond humanism at all. We are not post-man, postanthropos, posthuman.

Let's spell this out more clearly. What comes after postmodernism and poststructuralism is *anthropocenity*, the condition of living in the Anthropocene. This new time of "anthropos" is therefore not a *posthumanity* at all, genuine or otherwise. It is the late arrival of *humanism itself*, arguably for the first time. This idea is loosely echoed by Peter Frase in *Four Futures: Life After Capitalism* (2016), where he observes that

> Some leftist ecologists are suspicious of this term ["Anthropocene"], viewing it as a way of blaming ecological damage on humans in general rather than on capitalists in particular. But it doesn't have to be that; the Anthropocene can simply be a recognition that ecology must always revolve around human concerns. (2016, 104)

That's well put – and ironic. For it's only in the Anthropocene that humanism achieves its centuries-long quest to put human beings at the centre of all things. It's only in the present that human beings are understood to have radically impacted and changed the earth itself.

This is not an endeavor that humans have ever found themselves engaged in before. I mean, this is really drama at the scale of allegory and parable, but it is real. And we are those actors. We are those gods. And yet we are behaving as though the story is unfolding completely out of our control. (David Wallace-Wells, in Brady, 2018)

Today we are curators of this humanist reality – we evil geniuses. Obviously our unpredictable, chaotic present is not the future of mastery and control that Kant predicted of rational subjects. Nor is it the realization of Hegel's "end of history" as realized in knowing, at last, that "the real is the rational." It is perhaps more accurate to say that the real is the unrational, that the fantasy of humanism has finally become the nightmare reality of our time. For at long last humanism lives. In this sense, "History" can finally be said to have begun.

In short, the Anthropocene condition names the future we most surely deserve. And I'm afraid there's nothing "posthuman" about it at all.

16 Chthulucene?

In my view, the twenty-first century is having its own *Blue Marble* moment in the analog that is anthropocenity – the consciousness-raising realization that human beings have modified the entire Earth, mostly for the worst (from the perspective of actually existing life), and that its future is in large part determined by the activities of our recent past. Or again, our tradition of enlightened humanism, and with it the rational subject, has shaped the destiny of our physical world today. In this sense, the idea of the Anthropocene condition is as important today as the image of the whole Earth from space was to people living in the late 1960s and early 1970s. For in the Anthropocene, the *Blue Marble* has finally been elevated to a global philosophical concept. As such, consciousness is enlarged.

And so while a cultural theorist like Donna Haraway eschews the term Anthropocene, precisely because it reaffirms old fashioned humanist thinking, I recommend the opposite: doubling down on the fatality that is humanism. For if there's a "truth" of this new condition, it's that anthropocenity is the concrete manifestation of an old ideology about the subject that goes back to Western Enlightenment: the imperial subject rendered obnoxiously material, unavoidable, global.

> **It feels as though we are living through some weird perversion of the Enlightenment dream. Instead of humanity rationally governing the world and itself, we are at the mercy of monsters that we have created.**
> (Dale Jamieson 2014, 1)

This regrettably totalizing result, and the outdated privilege it represents, simply names the logical consequences of the Western worldview. It names the culprit and cause of the mess we are in: "Man." That's why Haraway's recourse to "tentacular" thinking and multispecies diversity is one viable way forward, while her alternative word for our epoch, the so-called "Chthulucene," is just another creative dead end. For while the former grasps at a future beyond the worst features of humanist thinking, the latter simply repackages the ambitions and limitations of a tone-deaf rhetoric floated in the 1980s.

The truth is, the word "Anthropocene" has captured the imaginations of people everywhere. Its sudden ubiquity is not because the word rebrands or calcifies our supposed dupedness as subjects of neoliberal capitalism (the "Capitalocene"). It is, on the contrary, a positive indication that Westerners have finally begun the hard process of taking *responsibility* for the mess we created since industrialization. The logic of *anthropos* has finally become the toxic present.

Of course Haraway, like Wolfe, is right to argue that we need a new language, new way of thinking, and new way of being human in this new world. But she is wrong if she thinks we can bury the problematic past by eschewing the word "Anthropocene." Language, including the metaphoricity of language, cannot really get us outside of language. There is no escape. There is nothing outside this Western anthropos; Western globalization has encircled, and has thereby enstrangled, the earth. And so it is that sometimes a word, in this case Haraway's alternative word "Chthulucene," performs the magic trick of signifying nothing beyond itself or, worse, effaces the very history that got us into this mess.

HARAWAY ON "ANTHROPOS"

I have an allergy to the particular etymology of the anthropic: the one who looks up, the one who is not of the earth, the one whose feet are in the mud but his eyes are in the sky; the retelling, once again, of stories that I think have done us dirt in Western cultures.

That all made me think. If we can have only one word, let's use *capitalocene*. But of course the fact is that we need more than one word. Capitalocene is a term I thought I had invented, but it turns out I most certainly did not. (Donna Haraway 2016a)

So forget Haraway. Our diverse and collective hopes turn on the possibilities of a truly radical shift in consciousness – a shift to new ways of being human and new ways of imagining ourselves. A shift toward what Chakrabarty, following Karl Jaspers, calls an "epochal consciousness" (2015). One thing is for certain: it can't happen via the overblown and ironically self-regarding, look-at-me ex-centric individualism of Haraway's dated wordism of thirty or forty years ago.

17 "Terraphilia: Earth Ethics"

In *Earth and World: Philosophy After the Apollo Missions* (2015), American philosopher Kelly Oliver takes a page from Cosgrove's and Lazier's discussions about the Apollo photographs to effect a confrontation with Kant, Arendt, Heidegger, and Derrida. The payoff comes in her last chapter devoted to "terraphilia," love of Earth, of which her pointed discussion of "earth ethics" is most useful.

CLIMATE ETHICS

The real climate challenge is ethical, and ethical considerations of justice, rights, welfare, virtue, political legitimacy, community and humanity's relationship to nature are at the heart of the policy decisions to be made. (Stephen M. Gardiner 2016)

Oliver argues that human beings, having evolved greater empathy toward animals, can similarly develop a greater empathy toward all the other cohabitants of Earth. So against the Enlightenment values of masculinist reason and recognition, she argues for the primacy of interspecies love; against an ethics based on self-interest and rational calculation, she argues for "response ability"; and against the ideals of disembodied, masterful subjectivity, she argues for "our embodied relationality" to others (234–6). In short, Oliver posits the value of eros against the Greek value of logos – reason and logical argumentation. For, as Arendt teaches, "*amor mundi*, or love of the world, can act as an antidote to the wordlessness of the desert of nihilism" (32).

Earth ethics provides an alternative to humanist disinterest in life, love, and existence. In this way Oliver repackages poststructuralist antihumanism, based on Nietzsche's attack on Western nihilism, as a theoretically coherent and ethically responsible way of confronting the environmental crisis today. She writes: "This ethos of the earth can provide the grounds for a nontotalizing, nonhomogenizing earth ethics, if we can imagine a dynamic ethics based on the response ability of biosociality and biodiversity rather than on universal moral principles that may close down the possibility of response" (240). In other words, Oliver heeds Lyotard's call from 1984 – no more slackening in the face of our attempts to dismantle truth, power, and reason: more "agonistics" – while invoking an openness to the Other found most insistently in the ethical philosophy of Emmanuel Levinas (cf. 197). From her own planetary perspective she essentially asks if we can "imagine the positive potential of global cosmopolitanism put into the service of the earth" (239), a cosmopolitanism restrained by mutual "dependence, belonging, and deeply shared bonds with those whom we may not even know exist" (229).

But if human beings can empathize with each other and with our animal friends, then perhaps we can empathize with Gaia itself. This is a position advanced by public intellectual Jeremy Rifkin in *The Empathic Civilization: The Race to Global Consciousness in a World in Crisis* of 2009 (a work that Oliver cites approvingly, if passingly, in her book). If so, empathy with Earth would be the umbrella under which an earth ethics can function. It's also the ethical content behind the idea that human consciousness is coming to know the world that is made by and that makes in its turn human beings. Or, stated differently, that behind the epistemological thesis that *knowing the world is experiencing and feeling the world* is an ethical thesis about how we *ought to be in this world*. If so then it's all about trying to extend these experiences to Earth, to Gaia.

It's worth qualifying that the way forward is not necessarily one that simply jettisons old-fashioned reason and recognition. As Foucault insists, we should reject the "blackmail" that forces an easy choice for or against Enlightenment (2007, 110). Empathy that acknowledges our relationship to Earth as a whole is not simply a repudiation of the Enlightenment tradition of reason and recognition. It's the realization that empathic identification with Earth alters the tradition of human "reason and recognition" that once set us apart from other animals. By the same token, human beings are *reasonable* to recognize themselves in the earth, this literal and figurative ground of all thinking. For, finally, the Anthropocene condition doesn't just supersede the past but builds on whatever features make sense for existence in a very different future.

So what about the two main traditions in contemporary Western philosophy? For some analytic thinkers, like Canadian philosopher Byron Williston (2015), the Anthropocene obliges us to reinvest in the old Enlightenment ideals of rational truth, justice, and hope. Williston argues that the Anthropocene is less "condition" than "project," one directed by ethical concern for future generations. His basic charge, fuelled by virtue ethics, is similarly bracing: we "need to become better people" (7; see Jamieson 2014, 186). For other thinkers of a more Continental bent, like Kelly Oliver, the solution lies along a path of rigorous antihumanist critique of Enlightenment ideals. But in the ways that actually matter, the differences between analytic and Continental approaches exist more at the level of style than substance. Both traditions interrogate the past, criticize the present, and aim for revolutionary ends, namely, a better future for life on earth.

Consequently I don't think we need to choose between analytic and Continental approaches to the climate crisis. In any case, this isn't the time for more sectarian debate about Western philosophical traditions, traditions that are just open toolboxes to help us imagine and create a better birdhouse tomorrow. Let's just get on with it.

18 Features of the Anthropocene Condition

That's enough about the "present." Before we turn to the future, let's gather up all thirteen features of the Anthropocene condition sketched so far, and add a fourteenth feature from "the future" (from section 3). Note that a final answer to the question of anthropocenity, built from the features summarized below, is provided one last time at the end of section 3.

1. The Anthropocene condition is a time of chaos and unpredictability, where the frequency of black swan events undermines the meaning of normalcy, common sense, and order.
2. Inequality exists everywhere in the world but will grow most dramatically within Western society.

3. Anthropocenity is a kind of gross literalization of Kant's insight into human reason and seems, as such, like a suitable foundation for realism. *The object world we know is a human world.*
4. Marxism no longer functions as a viable guide to the postcapitalist future, or for understanding the condition of life in the Anthropocene.
5. The masses feel almost universal disgust for politicians and, in turn, for democracy more generally. Hence the rise of populist rage.
6. Faced with the bankruptcy of baby boomer ideals in the rise of neoliberal globalization ("boomernomics"), anthropocenity reconnects society with the concrete realities of the Earth.
7. The hypocrisy of the rich has finally caught up to the sufferings of the poor.
8. The return of existential angst, only now on a planetary and species scale.
9. The irruption of gallows humour and giddy amusement, inevitable counterparts to existential angst about the future.
10. Individual and group psychologies in the Anthropocene flit between mania and depression.
11. The Anthropocene condition points to a revolution in *thinking*, to a shift in human *consciousness*, one that comes long after the Anthropocene proper was born. That revolution began in the late 1960s.
12. Anthropocenity names the debt incurred from at least a century of absolute freedom (and irresponsibility). That debt is a new form of absolute responsibility to the Earth.
13. Arguably the confrontation with human suffering (from climate change, the end of capitalism, population stresses, new technologies of automation) on a global scale is the last and also most important feature of anthropocenity. The "democracy of suffering" is the tragic foundation and common ground for social unity and social action in the future.

PART THREE

THE FUTURE
(CA 2008–2100),
ON THE
DEMOCRACY
OF SUFFERING

"Drive-By Looking Glass" (2018), Todd Dufresne

1 The Just Society

According to Plato, just souls thrive in just societies. And since there is only imperfect justice in everyday society, we inevitably find ourselves confronted with imperfect and unjust souls. The primary task of the ideal republic is, therefore, education on both levels of society and soul, the group and the individual. That education is reflected in Plato's commitment to purge the unjust society of vice while disciplining the body to obey the mind.

Plato's goal is to create better citizens by imagining not just a better society but a utopia of meritorious reason. To this end he argues that potential leaders are obliged to contemplate the eternal Forms, which are checks on the transient flotsam of everyday life. Plato's famous dialogue is actually, in this respect, an engagement with the Athens of his own time: how the decadent city state could be measured, known, judged, and improved by knowing the truth of *things* (like chairs and dogs) and the truth of *concepts* (like beauty and justice) (see Dufresne 2017).

On the one hand these ideal Forms are obviously ahistorical. Eternal and unchanging, they are the essence of *what is* in the world – ideals set apart from time and place. This is why Aristotle complains that Plato's idealism mistakes real men for "man," an abstraction, even as Plato fails to account for how stuff in the material world actually comes to be and change over time.

On the other hand, while Plato's philosophy may be ahistorical it's not antihistorical. For starters, his idealism (his metaphysics and epistemology) does not vitiate his pragmatism. It's always a mystery to new readers of Plato that he leavens his prescriptions about Being (Form) with the realpolitik of governance, arguing, for example, in favour of lying to citizens when necessary. These are the so-called "noble lies." But there's no mystery here. Plato tolerated lying because he knew that individual citizens aren't always just or smart and so aren't always equal to ideals; they may be less gold, in his analogy, than brass and iron. So while Plato's Forms are the sun around which human virtue rotates, he concedes that abstract ideals actually function more like models, like pretty paintings, of what is worth striving for in society. Of course *The Republic* is itself just such a painting, a tale best suited, Plato says, for "real men" – by which he means those ideal men who model justice for lesser mortals.

It's in this practical mode that Plato's thought remains historical and also proscriptive, in short, more than a philosopher's idealism.

2 Education

A just education is a moral matter with practical implications. In Plato's own life, *The Republic* marks what the classicist Eric Havelock calls "the introduction of the university system in the West" (1963, 15). In short, it marks out the intellectual and institutional territory of Plato's Academy. For Plato it is education, and by extension teachers, that drive morality and justice.

But Plato's investigation of a just education is also proscriptive, helping him imagine, for instance, the collapse of the biological distinction between men and women in favour of their myriad qualities of soul, namely their individual merits. Hence Plato's argument that, in a just society, women can perform any function – including the function of philosopher queen. In this sense Plato's philosophy, very nearly a species of science fiction, is a window onto a future radically different from his own present, a future inconceivable to the sensible Aristotle. Plato is a dreamer, in short, but clearly not a foolish one.

3 Whacking Stick

Plato's thought is characterized by its "essentialism," its violent reduction of reality to the universals of reason and mind. After Nietzsche, more than two thousand years later, this metaphysical project is dead or, if not dead, at least unbelievable. Not so his pragmatism. Plato's interest in everyday politics and morality survives the centuries, even if the discipline of philosophy has ceded this ground to the upstart social sciences like sociology, political science, and psychology or simply left it to its oldest enemies, those rhetoricians still at work in English and law departments.

Arguably it's essential to revive this ancient concern for the imperfect justice of everyday life, for politics in the broadest sense, since our time demands nothing less than a full reassessment of our present in light of our possible futures. My claim in this regard (proffered in section 1) is that politics in the Anthropocene can only be accomplished by employing the whacking stick of a better future.

In this sense Plato was right to look for a way to measure justice and on that basis establish models for how to govern the "city in high fever" that was his Athens. But he was wrong to establish Forms as the North Star of judicious judgment, just as Christians, following Plato, were wrong to do the same with God. Or rather, not "wrong" so much as not universally and eternally useful. Essentialism can't work – not only because human subjects are unequal to the ideal of divine knowledge. It's more simply the case that no thinking subject, after the thorough demystifications of the twentieth century, can possibly *believe* in it – especially as an ideal or model. Indeed, these ideals became their dialectical opposite, vehicles for power and oppression. Consequently Plato's ideals can no longer motivate and generate the justice he thinks we need.

Arguably the North Star must instead be the liveable possibilities of an ideal future – not securely known, quantifiable, and accessible to some sect of privileged thinker (a depth interpreter of some kind) but only guessed at, gestured toward, and open to the vagaries of human imagination. Our ideals must be conceivable and believable if they're to be motivational and achievable.

The trick is to think of the nascent future not only against the past, and not only in contestation with the fading neoliberal present, but as a call for creative openness to the myriad possibilities of what could be – of what is yet to come. Like Plato, in short, we need to seek out the *Kallipolis*, the "beautiful city," but one appropriate to our needs in this time and place.

4 Meaningful Gestures

Early in his fiercely intelligent book *After Nature* (2015), law professor Jedediah Purdy derides Roy Scranton's "self-important pronouncements" (4) about the end of civilization (recall that Scranton says it's "already dead") and our need to think existentially about our predicament (5). "This is just the sort of suggestive but, upon scrutiny, meaningless gesture," Purdy complains, "that makes talk of 'responsibility' feel self-important and ineffective" (5). Undoubtedly Scranton's thinking, like his prose, descends into near-absurdist ramblings (noted above) – and so Purdy's complaints are well taken. Scranton, moreover, inflates his rhetoric in the book version of his original essay, opening a final chapter with a purple flourish that would make Schopenhauer blush: "Death begins as soon as we are born. From our first moments in the world, blinking and crying in the light, we fly an unwavering arc to the grave" (89). For Purdy, the real work that needs to be done today is not existential – "This is just the sort of thing I want to avoid; I want this book to be worth reading" – but political, hence his sub-title "A Politics for the Anthropocene."

Of course, every discipline will engage anthropocenity in its own fashion. That's why discussions and solutions can be so predictable – elaborate realizations of the witticism that *every problem looks like a nail from the perspective of the hammer*. And so Scranton's *Learning to Die in the Anthropocene* elaborates a problem that makes broad sense to humanities scholars of a certain bent, while Purdy's *After Nature* elaborates a problem that makes sense to scholars of the law and politics. Which is fair enough.

Yet it must be admitted that Purdy's thinking about the responsible future isn't always so different from thinking that emphasizes the subjective and existential dimensions of the Anthropocene condition. In fact, Purdy's big claims on behalf of the determining role of imagination in the creation of human society are also made by Scranton. The basic claim is that imagination is productive or, if you prefer, generative. According to Purdy's view of it, Western society has created and then recreated nature, passing through the historical moments he calls providential, Romantic, utilitarian, and ecological. Each moment reflects an interpretation of the world, a way that human beings construct nature to suit our changing needs. Purdy's point is that we need to develop new, nonprecious ways of thinking about and relating to nature that make sense of our "Anthropocene condition" (4).

To this end Purdy comes close to rejecting the field of philosophy. "*Imagination* is less precise, less worked-out, more inclusive than *ideas*," Purdy says early on, "and it belongs to people in their lives, not to philosophers working out doctrines. Imagination is a way of seeing, a pattern of supposing how things must be" (22). Truly no one cares about a philosophy committed only to "doctrines" – including, perhaps most especially, those philosophers who remain oblivious of their own obliviousness. As gestures go, that kind of philosophy is a dead-end. Fine. But it's also a straw man. Moreover, law doesn't stand outside philosophy. Consequently it's in no position, literally and figuratively, to save imagination from elitist, sterile philosophy. Beyond that we can easily see, right from its beginnings with Plato, that the best philosophy has always been conceived between ideas and imagination and belongs, in Purdy's phrase, to "people in their lives." Philosophy belongs, in short, to human *existence* and is a major part of the ancient project to contemplate "how things must be."

It's not just that every Hegel inspires a Marx, someone dedicated to setting philosophy back on its feet. It's that thinking about ideas is also, and necessarily so, contextual and deeply political, even when it claims otherwise, and is therefore prone to generating unexpected effects. And that's because ideas, like imagination, belong to no discipline – even a historically and institutionally privileged discipline like philosophy. As Heidegger argues, thinking belongs to what is most precious, and most practical, about human existence more generally. And that kind of genuine, meditative thinking is by no means the same thing as "professional philosophy." Perhaps it rarely is.

5 Work of Imagination

Purdy privileges imagination because revolutions are formed under the influence of imagination. As he puts it, an act of "imagination is intensely practical" (7). That's true. But revolutions are also formed by good and bad *ideas* and *doctrines* – formed, in short, by myriad human beliefs. Consider Purdy's own example. The most recent "ecological revolution," the realization that the Earth is one complex interrelated system, is "deeply involved in contests over imagination, over the meaning of the world and the right way to live in it" (9). That's also true. But to be fair it's also what we usually mean by "philosophy." More precisely, it's what we mean when speaking of metaphysics (imagination), epistemology (meaning), and morality (right living). These contests become, in turn, the foundations of other disciplinary explorations that, as Purdy says, are "explicit and inescapable" in the Anthropocene condition. In effect, philosophy *in the best sense* helps us understand "how to live in a world that we cannot help transforming again and again" (9) – and it's bootless to render existential concerns of the kind Scranton discusses meaningless or useless.

Imagination is the human capacity to rethink and change our relationship to nature and, as Purdy insists, to remake it, responsibly and consciously. Above all he claims that we can no longer pretend that nature exists separate from us and from our ideologies and technologies. As he says toward his conclusion, "every technology will become part of the joint human–natural system in which we make and remake the world just by living here" (260). Paul Masson makes a similar point, echoing Marx's view of species being, writing that "We know the natural world only by interacting with it and transforming it: nature produces us that way" (2015, 245). The upshot is that we must abandon lingering romantic ideas about the naturalness of nature: ideas advanced most recently by the Pope (see, for example, #34, #117 2015). As Purdy says, we are now "post-natural," all "cyborgs in artificial worlds" (16). Nature is what we *make* of it, and what it makes of us in turn.

Human subjects are at the heart of this valuation. As Peter Frase puts it, "Any attempt to maintain climate, or ecosystems, or species is ultimately undertaken because it serves the needs and desires of humans, either to directly sustain us or to preserve features of the natural world that increase the quality of our lives" (2016, 103). Far from romanticizing nature or imagining a return to some pristine notion of nature, we must accept our roles as world makers. "We have no choice," in Frase's terms, "but to become more involved in consciously changing nature" (2016, 105).

ON NATURE

When the technologies that we have created end up having unforeseen and terrifying consequences – global warming, pollution, extinctions – we recoil in horror from them. Yet we cannot, nor should we, abandon nature now. We have no choice but to become more involved in consciously changing nature. (Peter Frase 2016, 105)

So instead of scary and elitist "ideas," the domain of philosophy, Purdy argues on behalf of "imagination," "styles," and "pictures of the natural world" (2015, 25–6), the domain of any thinking person. It's a spurious distinction, but it resonates well in our dumbed-down, opinion-based, highly mediated *Zeitgeist*. Appeals to imagination are in fact everywhere today, easily transcending turf wars between disciplines. In truth its proper category is neither everyday thought nor academic philosophy but *science fiction* – a field that inevitably bleeds into every facet of academic scholarship, even as it defines the two poles of the Anthropocene condition today, utopia and dystopia, that can be found throughout Western culture.

6 Doom Prognostication: "I see dead people ..."

Doomsday scenarios practically comprise their own genre today. Certainly there has been no shortage of diagnosis and prognostication about catastrophic climate change in the West. Dreams of apocalypse are obviously sexier than dreams of utopia, both reflective of, and constituting these existential times.

We've been forewarned in popular media. Samuel Jackson's character, a kind of demented James Lovelock, tells Michael Caine's character in *Kingsmen: The Secret Service* (2014), "When you get a virus, you get a fever. That's the human body raising its core temperature to kill the virus. Planet earth works the same way. Global warming is the fever; mankind is the virus. We're making our planet sick. A cold is our only hope. If we don't reduce our population ourselves, there's only one of two ways this can go: the host kills the virus, or the virus kills the host."

We've been forewarned in literary and science fiction ("cli-fi" or climate fiction). Kim Stanley Robinson in *Sixty Days and Counting* (2007): "CQ: So soon the whole planet will be developed and modernized and we'll all be happy! Except for the fact that it would take eight Earths to support every human living at those levels of consumption! So we're screwed! JQ: Dad. PC: No, that's right. That's what we're seeing. The climate change and environmental collapse are hitting the limits. We're overshooting the carrying capacity of the planet ... CQ: Yes. PC: And yet capitalism continues to vampire its way around the globe, determined to remain unaware of the problem it's creating. Individuals in the system notice, but the system itself doesn't notice" (Green Earth 941).

We've been forewarned in the popular press. Scranton in *The New York Times* (2013): "I see water rising up to wash out lower Manhattan. I see food riots, hurricanes, and climate refugees. I see 82nd Airborne soldiers shooting looters. I see grid failure, wrecked harbors, Fukushima waste, and plagues. I see Baghdad. I see the Rockaways [a peninsula savaged by Hurricane Katrina on Long Island, Queens]. I see a strange, precarious world."

We've been forewarned in highbrow trade books. Klein in *This Changes Everything* (2014): "In wealthier nations, we will protect our major cities with costly seawalls and storm barriers while leaving vast areas of coastlines that are inhabited by poor and Indigenous people to the ravages of storms and rising seas ... [Our] governments will build ever more high-tech fortresses and adopt even more draconian anti-immigration laws ...

In short our culture will do what it is already doing, only with more brutality and barbarism, because that is what our system is built to do" (49).

We've been forewarned in academe. Berardi in an interview with Andrew Pendakis in 2012: "War and violence will explode everywhere in Europe in the coming years. As the price of energy becomes unsustainable, relations with the Middle East and with Russia are destined to fuel war. Nazism is the destiny of Europe" (in Pendakis 2013, 172).

We've been forewarned in the scientific report. "The Intergovernmental Panel on Climate Change" in 2014: "Continued emission of greenhouse gases will cause further warming and long-lasting changes in all components of the climate system, increasing the likelihood of severe, pervasive and irreversible impacts for people and ecosystems."

We've been forewarned in the military report. The Pentagon's "Implications for US National Security of Anticipated Climate Change" in 2016: "*Now*, the effects resulting from changing trends in extreme weather events suggest that climate-related disruptions are under way. Over the *next five years*, the security risks for the United States linked to climate change will arise primarily from distinct extreme weather events and from the exacerbation of currently strained conditions, like water shortages. Over the *next 20 years*, in addition to increasingly disruptive extreme weather events, the projected effects of climate change will play out in the combination of multiple weather disturbances with broader, systemic changes, including the effects of sea level rise" (2016, 3).

We've even been forewarned in the investment report. The World Bank Group's "High and Dry: Climate Change, Water, and the Economy" in 2016: "Water-related climate risks cascade through food, energy, urban, and environmental systems. Growing populations, rising incomes, and expanding cities will converge upon a world where the demand for water rises exponentially, while supply becomes more erratic and uncertain. If current water management policies persist, and climate models prove correct, water scarcity will proliferate to regions where it currently does not exist, and will greatly worsen in regions where water is already scarce. Simultaneously, rainfall is projected to become more variable and less predictable, while warmer seas will fuel more violent floods and storm surges. Climate change will increase water-related shocks on top of already demanding trends in water use" (2016, vi).

UN SECRETARY GENERAL WARNING

Over the past decade, extreme weather and the health impact of burning fossil fuels have cost the American economy at least $240 billion a year. […] This cost will explode by 50 percent in the coming decade alone. By 2030, the loss of productivity caused by a hotter world could cost the global economy $2 trillion. (António Guterres, in Folley 2018)

So we've been warned – and then some. When citizens in the near future express alarm or surprise by a confrontation with the Anthropocene condition, it won't be because they haven't been told. It's because they refuse to believe it. That, too, will pass, because they are destined to experience the effects of this condition first hand, at the level of their material existence. Catastrophe written on the body is destined to become a basic condition of life in the Anthropocene. Back to this in a moment.

7 Eschatology is the New Black

It's important to realize that we aren't the first civilization to face imminent demise. Empires come and go, regimes rise and fall, entire peoples are vanquished and cruelly erased from history. The Indigenous peoples of North America, for example, live today amidst the ruins of their own culture and know something about endings. In parts of Canada, for instance, Indigenous people exist, not in some romanticized union with the land, but pauperized in communities resembling a "Third World" bubble within the surrounding "First": gated communities, only in reverse. The story is repeated around the world because the impacts of colonization and imperialism exist around the world. Another group keenly aware of endings is the unemployed working class of the American rust belt (see Hedges, 2012; Brian Alexander 2017). The collapse of industrialization has produced dozens of Red Zones, for example in places like Camden, NJ, Pine Ridge, SD, and much of Detroit. And these conditions, too, have been reproduced across the Western world (and beyond). Consequently there are plenty of models – cautionary tales – for what awaits us after the end of capitalism and the unfolding catastrophe that is global warming.

Even so, massive and traumatic change isn't automatically a bad thing over the long term. Major dislocations followed each of the Agricultural, Industrial, and Information revolutions, producing winners and losers, and yet few would advocate a return to the past. Industrialization, for instance, has been the most significant motor of social progress and freedom in history. This is, in part, Timothy Mitchell's argument in *Carbon Democracy* (2011): without oil wealth there could hardly have been the kind of democracy we enjoyed in the West for over a century. Of course if this claim is essentially and not incidentally true, causal and not just correlational, then the "end of oil" should make every freedom-loving person worry. What is freedom in the absence of oil? Is there more or less of it?

The Anthropocene condition presents us with an unhappy choice. Do nothing but hope for the best while expecting the worst, namely rampant inequality and fascism, or do something and hope for the best while mitigating our expectations of, and apathy toward, the worst. Perhaps we'll surprise ourselves. Human beings behave differently in a crisis, often more altruistically, as seen during times of natural disaster or war. Human beings can adapt when motivated.

One thing is certain. Humanity will either be broken or made better by the looming economic, climate, and population disasters. I propose we try to make it better and, to this end, consider a constellation of practical philosophical questions: *what better, how better, when better,* and *for whom better*? Let's come back to them in a moment.

8 Petroculture & the "Crisis of Desire"

In August of 2015 a group of mostly Canadian scholars and artists, led by the cultural studies scholar Imre Szeman, met in Edmonton to debate the pending shift to a time "after oil." In its aftermath the Petrocultures Research Group produced a little book, almost a report, called *After Oil* (2016). In it they lay out the need for thinkers to provide the mostly missing social and political analyses to help spur "planned transition" to a future we might actually want.

Arrows, courtesy of Pixabay

"Petrocultures Research Group" (2015), Imre Szeman

To this end the book puts paid to the idea that the arts have nothing to add when it comes to addressing our hegemonic petroculture. "While scientists may have definitely told us about the reality of global warming," they say in the conclusion, "they've given us no clue as to the path forward from the present to the energy futures we want" (73). And of course they're right. Social and political analyses are beyond the mandate or talents of many (but not all) scientists, and more scholars from the arts need to push their way into a field that remains naïvely under-theorized.

In "Triggering Transition," Szeman's group poses the inevitable question of "what comes after oil" (40) and in the following chapter provides the six dominant narratives currently in circulation. But the metanarrative that drives their thinking, and is more noteworthy yet, is a prescriptive one: the future *should* be more just and fair, and to this end it *should* be socialist or communist (they don't say which). To achieve this goal, Szeman's group wants to deploy conscious intentionality, forethought, and agency against the "impasse" of a problem, global warming, that seems not just beyond our ability to change but beyond our willingness to even try (18–19; 70). Such is our "default resignation" – as exemplified by Lovelock and Scranton – that is tantamount to a paralyzing nihilism. They write:

SIX NARRATIVES OF CHANGE

1. Transition from Below

2. Transition without Loss

3. Transition through Localization

4. Transition after Capitalism

5. Transition through State Reform

6. Transition through Catastrophe

(Petrocultures Research Group 2016, 29–30)

The default position is a disabling one. It is to assume that this transition is a purely technocratic problem that will be resolved through technocratic solutions. Such a position assumes that responsibility can be entrusted and handed off to someone else. (19)

In short, the analysis of petroculture is an attempt by arts scholars to stop passing the buck (to scientists, primarily, in the hope of a technological magic bullet like "geoengineering") and to take responsibility for our situation. It's obviously not because the arts "foretell the future, but because they open us to a thoughtful and responsible composure towards its uncertainties and possibilities" (24, 58).

The group sketches six "principles of intentional transition." They are agency and mobilization; collective stewardship of energy resources; equality; ethics of energy use; sustainability of energy; and rethinking the dogma of growth and development (25–7). Like Purdy, they are focused on the pragmatics of change. Just as tantalizing is their claim that the end of oil has or will trigger "a crisis of desire" – namely, a crisis of the consuming subject. If there is a bottom line in *After Oil*, it is precisely this looming crisis of *being* in the Western world. For if agency means anything it has to mean it at the level of the human subject – and so we're back again, not only to a description of the world but to a prescription of the way it should be and perhaps will be for us – for human beings.

What does it mean, once again, to be alive in a time of radical crisis and change? To be a subject of, and to be subject to, a life on the edge of catastrophe? One answer: subjects of the Anthropocene condition are subjects in flux but with none of the giddy, celebratory excitement of the postmodern subject. Desire in the Anthropocene condition therefore requires a worthy object beyond itself. Beyond narcissism. Beyond a pickled, mummified existence. And beyond our death-obsessed culture. Of course, a life well lived, in meaningful connection to others, is a tall order – perhaps the tallest.

Perhaps we can take solace in the fact that curating one's life – the tangible effect of human agency and desire – will be clarified, even simplified, as the conditions for existence get harder. The democracy of suffering practically guarantees it.

Utopia, courtesy of Pixabay

9 Inventing the Future

Prescription also appears in Nick Srnicek and Alex William's important and influential book, *Inventing the Future* (2015), but even more transparently. "Our argument," they say, "relies largely on a normative claim rather than a descriptive one" (112). Addressing the concurrent crises of capitalism, progressive politics, automation, and the environment, they don't just *describe* the world after oil but unapologetically *demand* it. Their approach diverges refreshingly from traditional progressive politics, taking inspiration from the organizational logic of the Mont Pelerin Society (MPS) – the group founded by Friedrich Hayek in 1945 to consciously contest the common sense of the time, namely the Keynesian economics of New Deal society. Why bother echoing the MPS, which after all was a conservative organization? Because Srnicek and Williams know a winning formula when they see it, just as Klein does when she adopts Friedman's "shock doctrine" to her own ends. From utter obscurity on the fringes of debate at the time, the MPS doggedly set up the intellectual foundation and practical alliances that, decades later, provided meaningful and coherent alternatives to the collapsing Keynesian policies of the early 1970s. In short, the MPS adopted a long-term strategy that made our neoliberal present possible.

It's worth emphasizing that, from the perspective of the immediate postwar period, this outcome (the neoliberal present) would have seemed highly implausible – a neoconservative daydream. But that's the lesson. Like Szeman's group, the authors insist that it's time progressives did some of their own daydreaming about the kind of future they want, and then take steps to make it happen.

For Srnicek and Williams, the old school "folk politics" of their progressive allies – grounded on unexamined commitments to localism, direct action, moral purity, identity politics, and grassroots protest, combined with a principled rejection of universal claims, the ideal of collective freedom, and modernity more generally – is a sure formula for continued failure in the face of the neoliberal consensus of our time. In this sense they have a coherent and utterly devastating explanation for the failures of the Left, and successes of the Right, over the last few decades.

The history of the Mont Pelerin Society demonstrates that what passes for political common sense can and does change over time, and progressives shouldn't be scared of utilizing the strategies and tactics that

"Machine Born" (2018), Mark Nisenholt

worked so brilliantly for establishing the neoliberal worldview. This is the same basic message advanced by Masson. Activists need to get their hands dirty through "engagement with the mainstream, involvement with political strategy, [and] an enduring structural project more concrete than 'another world is possible'" (2015, 262). For Masson this means the Left needs to embrace features once considered "the sole property of the right: willpower, confidence and design." For Srnicek and Williams this engagement cashes out as the need for new "platforms" based on exigencies of our time, among them a demand for full automation, the end of work, and the implementation of UBI – the universal basic income.

Perhaps because of their very explicit platform and demands, Srnicek and Williams consciously shift the question away from "what" to "who" comes after capitalism. In other words the hot question of the 1980s – who comes after the death of the subject? – becomes in *Inventing the Future* the question of who comes after the death of capitalism and its petroculture? Part of their answer is found in the subtitle, "Postcapitalism and a World Without Work."

10 Working, the Future

Since the question of the subject is, arguably, the essential question of the Anthropocene condition, the "postcapitalist" subject of Srnicek and Williams (and others) is similarly essential. For capitalism, built on an ideology of individualism and greed, helped make this new era possible. "Perhaps the most important question for building power," the authors argue, "is the question of who will be the active agent of a post-work project" (156). The traditional answer has long been some version of the "classical revolutionary subject" as envisaged by Marx and his followers. According to the best known of these narratives, this role is supposed to be played by the working class. But Srnicek and Williams concede that the proletariat, and with it the enabling condition of industrialized capitalism, has for decades ceased to be a revolutionary force of progressive change. In part, this is because increased automation has made factory work precarious and stripped of traditional points of leverage (for e.g., work stoppages along the line of production). And so they conclude:

Who, then, can be the transformative subject today? Despite the growing size of the surplus population and common immiseration of the proletariat, we must accept that no answer readily presents itself. [...] The fragmentation of groups of resistance ... means that the task today must be to knit together a new collective "we." (158)

But the authors are being coy, for in a way they do identify this emergent subject – or, rather, these subjects. This new "collective we" is what they call "the 'people' of populism" (160), a people likely to be "mobilized" by the kind of tangible demands, like UBI, that they (and many others) claim will knit together otherwise disparate stakeholders on both sides of the political spectrum.

Far from being a side issue of interest to philosophers alone, the old question of the subject simply refuses to go away in the Anthropocene. Toward the conclusion of their book, Srnicek and Williams return yet again to the emergence of a new subject after capitalism. "The pathway towards a postcapitalist society requires," they insist, "a shift away from the proletarianisation of humanity and towards a newly mutable subject" (180). The way forward is the kind of grand platform thinking they prescribe throughout the second half of their book: "This subject cannot be determined in advance; it can only be elaborated in the unfolding of practical and conceptual ramifications. There is no 'true' essence to humanity that can be discovered beyond our enmeshments in technological, natural and social webs." And again, "The postcapitalist subject would therefore not reveal an authentic self that had been obscured by capitalist social relations, but would instead unveil the space to create new modes of being." This subject is beyond essentialism, beyond the philosophy of Enlightenment rationalism. No longer will it be perfectly adapted to capitalism or, more broadly, to the constellation of features I'm calling boomernomics. So yes, there *is* an alternative to neoliberalism, but it must be forged by strategy, not just accepted by accident and fortune or left to the kind of righteous, judgemental, intolerant, bullying identity politics – "folk politics" – that have effectively destroyed (from within) progressive alternatives over the last few decades. In this respect and more, the aspirations, analyses, and criticisms of comedian Russell Brand are often more useful for progressive ideals than what you find with many intellectuals. For starters, he actually originates with, and so identifies with, the working classes that elite intellectuals often claim to represent (although, naturally, rarely as "organic intellectuals").

UNFUN PROGRESSIVES

I was hurt when a fellow protester piously said to me: "What you doing here? I've seen you, you work for MTV." I felt pretty embarrassed that my involvement was being questioned, in a manner that is all too common on the left. It's been said that: "The right seeks converts and the left seeks traitors." This moral superiority that is peculiar to the Left is a great impediment to momentum. It is also a right drag when you're trying to enjoy a riot.

Perhaps this is why there is currently no genuinely popular left-wing movement to counter Ukip, the EDL and the Tea Party; for an ideology that is defined by inclusiveness, socialism has become in practice quite exclusive. Plus a bit too serious, too much up its own fundament and not enough fun. (Russell Brand 2013)

The dream of freedom imagined by Marx and many other radicals still matters. People are right to demand more of it and to this end must embrace a future that has been stymied by capitalism. The best we can do, like the Mont Pelerin radicals, is take small practical steps in a direction that will probably surprise us all. Those steps are necessarily built out of the realities left over by boomernomics – such as the decline of work as made possible by the "second machine age" (see Brynnjolfsson and McAfee 2014). So instead of hanging our politics on the losing cause of unionized manual labour, which, after all, was nearly always white and male, and instead of fetishizing old fashioned notions about the essential role of work in human nature (as many leftists do, including the Pope [#128 2015]), Srnicek and Williams make a virtue of necessity and hang human freedom on the winning cause of a nonwork future for all.

Of course mass unemployment will become a condition of greater freedom *or* greater instability; social justice *or* fascist repression; widespread satisfaction *or* revolutionary dissatisfaction and violence. As binaries go, the choice is stark – and, therefore, not much of a choice at all. It's tantamount to a choice between order and chaos, the possibility of happiness or the unavoidability of misery.

The Anthropocene condition is this aspiration for *what* and, most especially, *who* comes after the end of the impasse in which we now exist. Careful thinkers and tricksters, midwifes of this future, urge us to carefully, consciously prepare the way for it. In the meantime, we can already see that the subjects of the near future will have a new relationship to work and, by extension, to leisure. Mass unemployment and under-employment means that human beings will be obliged to deploy their desires in new ways. Let's make sure these new ways create the foundation for a new society and not become harbingers of its utter dissolution.

11 Rebellion & Revolution

Srnicek and Williams go further to characterize this freedom of the future as gleaned from the aspirational literature. "This freedom finds many different modes of expression," they say, "including economic and political ones, experiments with sexuality and reproductive structures, and the creation of new desires, expanded aesthetic capabilities, new forms of thought and reasoning, and ultimately entirely new modes of being human" (180–1). That's all reassuring. But in truth they don't spell out very carefully what will happen if mass unemployment is *not* offset by something like the Universal Basic Income – at least beyond the example of that one great mass of unemployed Americans, the unemployable prisoners, hard at work in private prisons (and whose labour is really a pernicious echo of institutional slavery).

CARCERAL CAPITAL

At least 37 states have legalized the contracting of prison labor by private corporations that mount their operations inside state prisons. The list of such companies contains the cream of US corporate society: IBM, Boeing, Motorola, Microsoft, AT&T, Wireless, Texas Instrument, Dell, Compaq, Honeywell, Hewlett-Packard, Nortel, Lucent Technologies, 3Com, Intel, Northern Telecom, TWA, Nordstrom's, Revlon, Macy's, Pierre Cardin, Target Stores, and many more. All of these businesses are excited about the economic boom generation by prison labor. Just between 1980 and 1994, profits went up from $392 million to $1.31 billion. Inmates in state penitentiaries generally receive the minimum wage for their work, but not all; in Colorado, they get about $2 per hour, well under the minimum. (Peláez 2004)

But perhaps they don't need to spell things out because the scary alternative to UBI in particular and social welfare more generally is so obvious. Set aside concerns that UBI is a conservative plot to gut social programs, since that makes UBI a symptom of decline instead of recovery and revolution. Set aside cynicism. Without a practical and universal mechanism to placate millions of newly unemployed workers – without something like the UBI – the future will be consumed by chaos. It will be a future of would-be incarcerated citizens, victims of disaster capitalism after the "end of capitalism." Worst of all, the chaos could degenerate into "exterminism," the purposeful extermination of "surplus populations" that contribute nothing to the advancement of the elites (see Frase 2016, 124–30). Automation, therefore, as harbinger of genocide. As the Marxist geographer David Harvey puts it, "Deaths from starvation of exposed and vulnerable populations and massive habitat destruction will not necessarily trouble capital (unless it provokes rebellion and revolution) precisely because much of the world's population has become redundant and disposable anyway"

Prison fencing, courtesy of Pixabay

(2014, 249; see 264). That's horrifying, more so because he's not wrong. But everything hangs on Harvey's bracketed qualifier – "unless it provokes rebellion and revolution" – which at some point I take to be a given.

> [I]t's impossible to say now if neoliberal ideology will fade, and give way to a return to social solidarity, or if the criminal class that has grown up in the shadow of neoliberal deregulation will instigate ethnic and national war, launching a planetary genocide for the possession of decreasing resources. (Franco Berardi 2011, 124)

Life in the Anthropocene is not, and probably won't be, catastrophic for everyone. But it certainly will impact everyone, with only one or two degrees of separation between the lucky and unlucky. While there have already been untold deaths around the world from climate change, there is still a modicum of order. As Heidegger would have it, everything still functions – more or less. But dysfunction looms and so too does a collectivist future based on rebellion and revolution.

12 Calm Before the Storm, a.k.a. the "Interregnum"

Szeman's group wonders aloud if "[p]erhaps we are already [living] in the *after*" (Szeman et al, 2016, 61). That's a fair and reasonable observation – and one I've implied by virtue of my dating scheme (a "future" beginning in 2008). But if Wolfgang Streeck, director of the Max Planck Institute for Social Research in Cologne, is right, then the future really hasn't quite arrived. In *How Will Capitalism End?* (2016), Streeck agrees that capitalism is dead but not quite buried, arguing that we exist today in an "interregnum" between what we knew and what is yet to come. Time itself seems to have paused as History or, more prosaically, the educated public, hold their breath. Streeck calls it capitalism in "limbo," a "prolonged period of social entropy" (13). This makes sense.

This interregnum is a reflection of the fact that there are no viable collectivist movements, or no alternative elite, waiting in the wings to take over as capitalism increasingly fails to deliver freedom, subsistence, and security to the masses. This is precisely Srnicek and Williams's complaint: there is no progressive version of the Mont Pelerin Society ready to capitalize on the failures of neoliberal capitalism. There's just self-congratulatory, masturbatory folk politics – a politics of pure style, or perhaps just vanity, where man-buns, tattoos, and eyelash extensions; correct beliefs about animals, cars, clothes, food, movies, music, and politics; obsessive and haughty indignation over microaggressions; and, just a tweet away, where virtue signalling and call-out culture function as weak substitutes for engagement with the eclipsed dreams of freedom and revolution. Meaningful, impactful citizenship is thereby gutted even as one purges strategic allies from one's bubble of correct thinking and, in its own way, of conformity. In other words this new puritanism is fuelled, ironically and despite legitimate concerns about (1) marginalized identities and (2) the loathsome actions of alt-right bigots, by intolerance toward difference. The prosaic expression of such jejeune self-regard is a wide-eyed abrogation of agency and corresponding denial of moral complicity in the wider world.

MORRISON ON THE PEACOCK

Too much tail. All that jewelry weighs it down. Like vanity. Can't nobody fly with all that shit. Wanna fly, you got to give up the shit that weighs you down.
(Toni Morrison 1977, 179)

Streeck characterizes the collection of citizens living through this inter-regnum period as "less than a society," a "*post-social society,*" "de-socialized capitalism," and "*society lite,*" one characterized by declining institutions and a culture marked by uncertainty, indeterminacy, and individual luck (12–13, 38). The elite will be inclined to market this perilous condition as a kind of radical freedom, doubling down on the individualism that's so problematic for the planet's future. To the extent that elites succeed in this sleight of hand, Streeck expects to see the rise of four behaviours – "*coping, hoping, doping,* and *shopping*" – which he organizes under the umbrella term of "*competitive hedonism*" (41, 45; his italics). In short he expects more false consciousness. More spa weekends, better winter tans, funkier hair dyes, cooler piercings, tighter abs. Ultimately "collective resignation" will be "the last remaining pillar of the capitalist social order, or disorder" (15).

But that's capitalism. Arguably Streeck doesn't pay enough attention to the other dire problems facing humanity. For we are facing catastrophes that will test "limbo capitalism," what Yanis Varoufakis calls "bankruptocracy," the "rule by bankrupted banks" (2011), beyond its capacity to spin false consciousness as radical freedom. This is Masson's basic point in *Postcapitalism*. Perhaps capitalist elites can hang on for a long time but not with the population and climate issues about to explode over the next few decades and not, moreover, in the wake of the crisis of humanist reason that has marked twentieth-century thought more generally. So yes, we are in an interregnum, and sure, competitive hedonism is the kind of marketing strategy we've grown to expect from our elites (and to enjoy as consumers, especially in the West). But no, Gaia waits for no one.

So Naomi Klein is right: massive change is imminent. If we do nothing at all about climate change, then everything will change – and soon. Climate events will be frequent and catastrophic, politics will degenerate into autocracy, food and water insecurity will reign, disease will spread, war will erupt everywhere, and hundreds of millions of people will suffer and die. If capitalism survives it will be some form of absurdist disaster capitalism based on security and selective rebuilding, not competitive hedonism. Based, in short, on chaos and war. As the Marxist scholar Samir Amin says, "we cannot discuss how to prevent war because war and situations more chaotic are inscribed into the logic of this decaying system" (in Amin et. al. 2018). In my view it's possible to maintain this system for some time but not forever. It is simply unsustainable; there are too many balls in the air for capitalism to juggle, including an angry population. We have, in short, reached what some critics are calling (by analogy with "peak oil") "peak capitalism." That said, we are probably years beyond peak capitalism – as we all wait, as though playing musical chairs, for the next global recession to drop.

If, on the other hand, we do something serious about climate change, then, once again, everything will change. For it means that the lingering shadow of capitalism has to go. In both scenarios, old expectations around growth, development, work, consumption, and accumulation will crumble. This means, ultimately, that our societies will no longer be embedded in the economy but that our economy will be embedded in our societies.

In other words, even the best scenario foretells of catastrophic events, since the grandest actions today won't stop a dangerous rise in climate, a rise that is guaranteed by our long-accumulating levels of existing carbon emissions. But a three-degree rise in temperature, while catastrophic, is far better than a six-degree rise. No doubt, nonlinear climate-generated disaster and death, from protracted droughts to the spread of lime disease to war, will be either (1) very bad or (2) very fucking bad. Both scenarios flirt with mass extinction but only the scenario of a six-degree increase practically guarantees it will come to fruition within one hundred years and maybe sooner. We are making this future now; there's no "opt out" clause for anyone.

13 Optimistic Pessimism: The Case for "Transition through Catastrophe" & "Transition from Below"

Klein rightly advocates for hope and freedom over despair and enslavement. Recall again her missive for a "Marshall Plan for the earth" (458), a planetary shift that will force humanity to address "the unfinished business of liberation" (459). Only very big change will partially avert literal and moral catastrophe. As she says, global warming "changes everything." So yes, big change is inevitable. And sure, seismic change can occur quickly. "When fundamental change does come," Klein declares of history, "it's generally not in legislative dribs and drabs spread out evenly over decades. Rather it comes in spasms of rapid-fire lawmaking, with one breakthrough after another. The right calls this 'shock therapy'; the left calls it 'populism'" (461). But dribs and drabs will no longer suffice. Time is up. As Klein and her activist peers say in the "Leap Manifesto" of 2015, "small steps will no longer get us where we need to go" – hence the urgent need to "leap" forward (see also Battiston 2018).

Yet we must concede that the *fact* of climate change hasn't really changed human consciousness very much, at least not yet, so much so that a cynic might counter with an alternative slogan, "this changes nothing." This dark sentiment is, in fact, voiced by Berardi. Consider his remarks about antiglobalization activists protesting on the streets of Seattle from 30 November to 1 December 1999. For Berardi, although the activists may have achieved an "ethical consciousness," the movement itself "changed nothing in the daily life of the masses" (Berardi 2011, 12–13). "Ethical demonstrations," he insists, "did not change the reality of social domination," that is to say, did nothing to interrupt capitalist exploitation. People cocooned in Green Zone bubbles of wealth certainly remain wary of the ecological and economic futures but are largely unmoved. As the Slovenian philosopher Slavoj Žižek puts it, "we *know* the (ecological) catastrophe is possible, probable even, yet we do not *believe* it will really happen" (Žižek 2017).

Clearly education and reason (good schools, good universities) are necessary but insufficient causes of a shift in consciousness – most especially a shift that goes beyond (mere) ethical insight into our situation. This is the crux of a debate between Gardiner, the University of Washington philosopher, and David Weisbach, the Chicago law school professor and economist (Gardiner and Weisbach 2016). Gardiner's pleas for a renewed ethics in the face of climate change are countered by Weisbach's call for traditional cost-benefit analysis and feasibility. It seems, for Weisbach, that ethics *prescribes* but *doesn't change* anything. It's too utopian. Too ideal.

"Climate change," Dale Jamieson observes, "poses the world's largest collective action problem" (2014, 105; see Jamieson 1992, 151). That's exactly right. Arguably what's missing is "a common ground of understanding and a common action" of the sort that Berardi highlights in *After the Future* (2011, 14). What's missing, in short, is a trigger that can't be ignored, one that functions less like a sermon or lecture than like a slap to the face – or worse. Berardi, for one, can't imagine it. For him "collective intelligence" or "general intellect" is only theoretical (163): "I don't see any discernable subjectivation, resurrection of consciousness, or emancipatory forms in the foreseeable future" (158). He doesn't even see a place in the "psychotic" present for political action and political theory (see Berardi 2018).

But "the democracy of suffering" entails more than brooding fatalism. Let's follow its logic in the looming ecological crisis. When enough people have suffered the direct effects of climate change and know with an embodied certainty that such effects are caused by climate change, then we will have a trigger big enough to effect the kind of societal, political, economic, intellectual, and emotional changes that we obviously need. When rich Westerners themselves are directly impacted. When we are all obliged, by harsh contact with physical reality, to recognize the Anthropocene condition. When it's *too late* to avert catastrophic disasters everywhere.

THE GREENHOUSE EFFECT

"The Greenhouse Effect" (2015), Andy Singer

When we are forced to seriously ponder, with Scranton, that important parts of our civilization are dying. Then things will change. As Bill McKibben puts it, "Mother Nature is a very powerful educator" (in Ottesen 2019). This is an argument, in short, about winning by losing, a concession to the fatality, the inevitability, of catastrophic climate change.

Against Berardi – here is my hope – the democracy of suffering will transform the *globalization of indifference* into the *globalization of empathy*. And this transformation will indeed make possible a "common ground" for social action. That common ground includes the earth on which we live and die. In Chakrabarty's formula, anthropocenic climate change throws us "into the inhuman timelines of life and geology, [and] also takes us away from the homocentrism that divides us" (2015, 183). Yet "epochal consciousness" will always be on the side of human beings or, if you prefer, on the side of representation – the way we conceptualize and know the external world and, indeed, know each other. If we are newly formed "geological agents," it's precisely because of "a shared catastrophe" (2009, 218), namely the climate crisis. This experience is collective. So when Chakrabarty concludes, finally, that "There could be no phenomenology of us as a species" (220), as a whole, precisely because "one never experiences being a concept," he effectively repeats Husserl's mistaken claim of 1934. Just as *Earthrise* and *Blue Marble* explode Husserl's insight that human beings cannot (phenomenologically) experience an earth that moves around the sun, so too has the Anthropocene exploded the claim that human beings cannot (phenomenologically) experience what Chakrabarty calls "an emergent, new universal history of humans" (221). For the democracy of suffering ensures that climate catastrophe will indeed be experienced universally and concretely. As Chakrabarty says by way of concluding:

> Species may indeed be the name of a placeholder for an emergent, new universal history of humans that flashes up in the moment of the danger that is climate change. But we can never *understand* this universal. [...] Yet climate change poses for us a question of a human collectivity, an us, pointing to a figure of the universal that arises from a shared sense of catastrophe. It calls for a global approach to politics without the myth of a global identity, for, unlike a Hegelian universal, it cannot subsume particularities. We may provisionally call it a "negative universal history."

So-called species history or species thinking, the reorientation of human time to geological time in the Anthropocene condition, leads us to a we, an us, a collectivity that is made possible by the democracy of suffering. It may be a "negative universal history." Fine. But it's known because experienced, and known because the collapse of the object world is creating a mass subject brutalized by contact with external reality.

Of course the timing of this epochal form of consciousness, along with the newly found agency and capacity for empathy it implies, may seem belated – especially for nonhuman life on earth, for example, for the Bramble Cay rat of Australia and for many insects in the Amazon. The "recognition of the otherness of the planet" may be imperfect and therefore muted, depending on one's concrete experiences of the Anthropocene condition. And yes, the catastrophe won't be avoided. But it isn't necessarily too late for human civilization as a whole or, indeed, for species history and species thinking – for epochal consciousness. It isn't too late for those who survive the catastrophe to live lives of greater equality. And it isn't by any means the end of meaningful and joyful existence (see Jamieson 2014, 238). As Stanford historian Walter Scheidel (2017) argues, catastrophe has long been the royal road toward greater equality and social justice:

> Throughout history, only massive, violent shocks that upended the established order proved powerful enough to flatten disparities in income and wealth. They appeared in four different guises: mass-mobilization warfare, violent and transformative revolutions, state collapse, and catastrophic epidemics. Hundreds of millions perished in their wake, and by the time these crises had passed, the gap between rich and poor had shrunk.

So even if it's true that empathy for others must begin with personal experience, the *globalization of empathy* – based on catastrophic climate change and the emergence of epochal consciousness – is nonetheless a step forward for human civilization. It's also a path beyond this dark moment.

DAVID WALLACE-WELLS ON THE USES OF CATASTROPHE

I think extreme weather is really useful in this respect, too. It used to be that if you were trying to get people scared about climate change, you'd have to point to some distant future. Now there's so much horrifying climate disaster going on that you can just point to the news, to meteorological data, and most people understand already that we're living in an unprecedented time. I think that's very helpful, even if it's also terrifying. (In Brady 2018)

Al Gore likes to say that "watching the weather channel is like reading from the Book of Revelation." Of course, even a sexy pitch like this loses its shine with every retelling. People tire of doom and gloom. Human psychology, we are often told, despairs in the face of imminent catastrophe. But watching and reading has nothing on *living in* the Book of Revelation. Climate change won't always be a spectator's sport.

In the meantime we are very obviously drifting toward a barely intended global fascism led by elites but supported by middle and upper middle class folks like me – all desperate to hang onto money, power, and things. "Drifting" because neoliberal capitalism has at last exhausted all avenues of capital, leaving an undemocratic and nasty status quo that no one likes but nonetheless accepts, in Fisher's phrase, as our inescapable "capitalist realism." We are all proverbial frogs in a pot of boiling water, totally unable to make the necessary changes. But the truth is we already know what's required. The iconic Canadian professor and science journalist David Suzuki spells it out explicitly in his "Carbon Manifesto" of 2013. Here are his five demands: fossil fuels must stay in the ground; natural carbon sinks, like oceans and boreal forests, must be protected; we must switch to renewables within one generation; we need a carbon tax now; and scientists must be free from political interference to communicate their results. These are obviously good and sensible demands. Similar demands are echoed by many other informed activists.

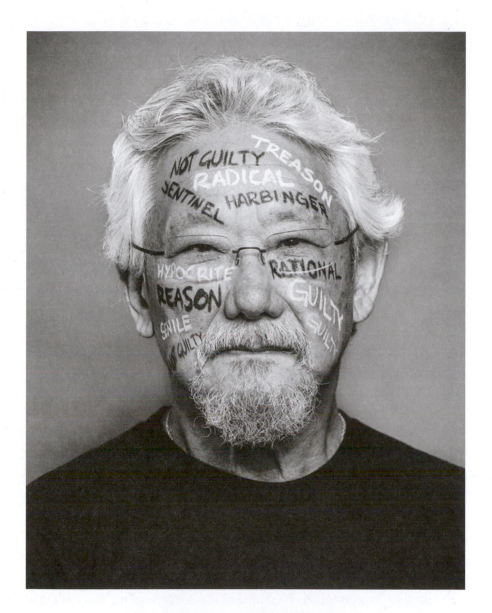

David Suzuki. Permission kindly given
by David R. Donnelly and Laurie Brown.

"GOOD NEWS" AT 350.org

1   We know exactly what we have to do – keep fossil fuels in the ground and quickly transition to 100% renewable energy.

2   Renewable energy is getting cheaper and more popular every day. In fact, global carbon emissions have already started to slow due to the rapid growth of clean energy.

3   We're not alone – the worldwide movement to stop climate change and resist the fossil fuel industry is growing stronger every day.

And yet we drag our feet, in large part because capitalism functions, again as Fisher argues, as "a kind of invisible barrier constraining thought and action" (2009: 16). Changing the status quo by contesting the remains of capitalist realism is hard. Intentional transition to socialism, communism, or some kind of surprising collectivism is hard. Decarbonizing our economy and culture is hard. Responsibility is hard. Accountability is hard. Yet so-called "radical" consciousness of the Suzuki variety is finally the most "realistic" consciousness for us today – since it represents the interests, not just of regular people but of existence itself. Of survival.

Qualified optimism is justified. Capitalist realism can't and won't last. The future will be driven by a consciousness committed to collectivist politics, a populist "we" or "us" – and by some version of a command economy. "For nearly two centuries," the philosopher Alain Badiou writes in *Le Monde*, "we have known that we must begin the new age, one of incredible technologies for all, tasks distributed equally to all, the sharing of everything, and the educational affirmation of the genius of everyone" (2018). Badiou calls it "communism," the name of the collective future as imagined by the recent past. But it really doesn't matter what it's called. What matters is that, in time, humanity will upend the apathy, despair, and "ecological grief" that fuels the self-fulfilling prophecy of mass extinction.

OPTIMISM

The people of the world will simply not allow
economic elites, with the help of the state, to
maintain toll-gates between us and the technologies
of abundance, skimming off most of the benefits in
return for letting some of them trickle down.
(Kevin Carson 2014)

In "A Call to Arms," American activist and founder of 350.org Bill McKibben argues that a "loud movement – one that gives our 'leaders' permission to actually lead, and then scares them into doing so – is the only hope of upending that [self-fulfilling] prophecy" (McKibben 2014). That's certainly inspiring. The people will rise up and politicians will be scared. But positive change or a "leap forward" to a new age won't happen without a push – without catastrophic climate change, the democracy of suffering, and a global shift in human consciousness.

So there's hope. But it's a desperate and perverse sort of hope. Unfortunately, optimism isn't always the opposite of pessimism.

14 The Democracy of Suffering

Not surprisingly, the legal scholar Purdy is contemptuous of the claims made on behalf of a shift in consciousness. It can't be wrong, he says, "But is it helpful?" (2015, 260). His scepticism echoes the derision he shows for Scranton's style of engagement. Against it he reaffirms, also unsurprisingly, the utility of his own field. "Some environmentally beneficial changes can follow from shifts in consciousness," Purdy sniffs, "but the biggest material changes happen through changes in the legal and economic infrastructure that guide human energies and activity" (261). Of course he's right. But no one advocating a shift in consciousness (or, indeed, advocating for the usefulness of imagination or of ethics) denies the concrete, practical benefits that will flow from such a shift. It's not about kumbayas and empty platitudes but about the kinds of practical measures that are Purdy's métier. Changes to regulations and changes to the way we organize the economy are what a change of consciousness heralds. For once again *we are the problem*, a refrain that functions like a mantra in Scranton's book. In fact, this claim is repeated by the pragmatic Purdy: "The lesson of the past fifty years is that humanity itself is the challenge" (159). And so there we have it once again. Scranton and Purdy, dreamer and pragmatist, share common ground.

Arguably the confrontation with the inevitability of human suffering on a global scale – where the old terrors of the global South have finally caught up with the citizens of the global North and then some – is the most important feature associated with the Anthropocene condition. For the democracy of suffering has the power to do to human consciousness what Apollo 17 did in December 1972, only now at the material level of our flesh, our bodies. Just as humankind saw the *Blue Marble* for the first time and understood at once, with a photograph, the fragility, beauty, and singularity of the planet Earth; just as people were thereupon motivated to start modern environmentalism and movements like the *Whole Earth Catalog*; so too will mass suffering in the Anthropocene make people understand the fragility, beauty, and singularity of human beings and human civilization. For his part, Scranton deploys an old nugget, philosophy-as-preparation-for-dying, to prepare for and imagine a world after the Holocene. Which is fair enough. But obviously individuals won't need philosophy proper to know, in a deeply personal way, the experience of suffering. Suffering will almost certainly gather survivors together to create meaningful new collectivities. Consciousness will be therefore raised, or at least altered, because the very content of consciousness will have been altered. It will reflect back upon and, moreover, *know* this hostile external world. In other words, *mass suffering* will accomplish what traditional philosophy never could, and it will accomplish what *mass labour* never could in the Marxist schema. It will collapse the difference between subjects and objects.

In his famous "master/slave dialectic," Hegel carefully qualifies the obvious: there is no lesson learnt by a "struggle to the death" when the subject actually dies. The dead are beyond a redemptive shift in consciousness. At the same time, it's worth stressing that survival alone doesn't guarantee epiphanies or, by extension, philosophy. As Plato understood perfectly well, people aren't always or even often built for the love of wisdom, for thinking, for reflection. It's too hard and is seemingly too removed from the object world. Not so with suffering. The experience of suffering, like the experience of death, makes everyone a *kind* of philosopher – which is probably all that Scranton has in mind.

Or is that rank sentimentality? As the old Freud once quipped while observing Jofi, his Chow dog: "No one looks more philosophical than a dog gnawing on his bone." And, really, that's the crux of it. Will the experiences of the Anthropocene condition make humanity better, kinder, more generous, wise? Or will it just see us divided once again into tribes, meaner and stupider than ever, the better to gnaw on the bones, not of our lost civilization, which is still a grand abstraction, but of our own personal losses, trials, and tribulations? Will grief and personal trauma trump wisdom? Not even the democracy of suffering can answer this question with perfect confidence.

15 Practico-Philosophical Questions

In response to the claim that we must try to build a better world, let's get back to our four practico-philosophical questions. In what way should we strive to make our evolving society better – and how? This is easy. What's obviously better is any form of conscious collectivism, since individualism as an ideology is dangerous and increasingly useless. This means a conscious end to rapacious forms of capitalism – which is to say "capitalism," full stop – and the end of the dogma of development and growth. It means, therefore, the end of work for the sake of working, certainly the end of what the anarchist anthropologist David Graeber calls "bullshit jobs."

"BULLSHIT JOBS"

Huge swathes of people, in Europe and North America in particular, spend their entire working lives performing tasks they secretly believe do not really need to be performed. The moral and spiritual damage that comes from this situation is profound. It is a scar across our collective soul. Yet virtually no one talks about it. (David Graeber 2013)

In this lattermost respect, automation of health, medicine, surgery, law, accounting, finance, programming, education, and much more, spells the end of work in our lifetimes for many people in the comfortable middle and upper middle classes. The situation is nicely captured by an article in *Wired* (11 February 2013) that declares, "IBM's Watson is better at diagnosing cancer than human doctors." Artificial intelligence has only gotten better since 2013 and is set to revolutionize "white-collar" work. This point is usefully summarized by Streeck:

> *Electronicization* will do to the middle class what mechanization has done to the working class, and it will do it much faster. The result will be unemployment in the order of 50 to 70 per cent by the middle of the century, hitting those who had hoped, by way of expensive education and disciplined job performance (in return for stagnant or declining wages), to escape the threat of redundancy attendant on the working classes. The benefits, meanwhile, will go to [what Collins calls] "a tiny capitalist class of robot owners" who will become immeasurably rich. (2016, 10; see also Collins 2013, 37–9; Brynjolfsson and McAfee 2014; Srnicek and Williams 2015; and Frase 2016)

In addition to the end of mechanized factory work, a fact of life for many over the last four decades, and in addition to the coming demise of well-paying jobs for managers in the information age, we also have the demise of jobs associated with the transportation of people and goods, since driverless vehicles (safer, more efficient, cheaper) will take over our streets and highways; the rise of car sharing and the concurrent decline of privately owned cars in major cities and thus the disappearance of car dealerships and sales positions; the demise of the fossil fuel industry that follows the decline of cars and, moreover, the broad shift to highly efficient and nearly free renewable energies everywhere; the continued shift from bricks-and-mortar stores, with all their associated costs, to warehouses for online purchasing of everything. Throw that all into the pot and one can begin to see that the work left over after the end of capitalism is the work required to maintain society and fulfill our own needs. Instead of lives spent doing soul-destroying and unfree labour, one might learn how to play the ocarina, tell a joke, or read a novel.

This conclusion isn't mere fantasy. Tectonic shifts in where we work and what we do in our jobs have been reported for years now. For example, in "The Talented Mr. Robot: The Impact of Automation on Canada's Workforce," the Brookfield Institute in Toronto reports that "nearly 42 percent of the Canadian labour force is at a high risk of being affected by automation in the next decade or two" (Lamb 2016, 5). Similar findings hold for the United States. In "The Future of Employment" two Oxford professors conclude that the "probability of computerization" in some fields is high enough to warrant concern. "According to our estimate," the authors write, "47 percent of total US employment is in the high risk category, meaning that associated occupations [e.g., sales, service, and transportation sectors] are potentially automatable over some unspecified number of years, perhaps a decade or two" (Frey and Osborne 2013, 38). Less well appreciated is how utterly transformative automation, married to evermore sophisticated artificial intelligence, will be on the various professions – and how quickly it will remake the economies of the West.

But let's get back to the "better world" scenario. It's clear that subjects living in the Anthropocene are destined to be unemployed subjects or, better put, alternatively deployed subjects. Assume the best outcome for this future of automation. Instead of our default identities as consuming subjects, we will of necessity become more responsible and self-aware subjects but, also, surely, more inclined to become meaning-seeking subjects. In the broadest sense of the word, many of us will become thinking or philosophical subjects. True, that will be burdensome for some, and possibly tiresome for everyone at times – like the early days of mandatory recycling. But it will also be more satisfying, more meaningful, since meaning derives from being part of something bigger than one's self.

In political terms, it means that governments will be highly centralized and thus able to plan for and respond quickly to serial crises, something impossible under capitalism and the dogma of invisible market forces. It means passing tough environmental laws and regulations, while setting ambitious energy targets. It means more state ownership of essential services, like energy and water, while rolling back privatization under neoliberalism. It means the end of austerity as a policy for downloading everything onto the vulnerable poor and the imposition of higher taxes on corporations and the rich. It means a commitment to collectivist welfare schemes for the increasing mass of unemployed and underemployed people, such as minimum wages and the implementation of a living wage through the UBI. It means free or nearly free education from childhood to adulthood, as both a practical and spiritual good. It means a commitment, primarily through

The future of employment

About half of today's jobs will likely be done by computers in a decade or two. Automation has so far taken over mostly well–defined routine tasks, shifting jobs from middle–income manufacturing to lower–income service jobs. As computers get better at for example perception - think self–driving cars - those services jobs are likely next up to be replaced by machines. Frey and Osborne (2013) estimate the probability of each job becoming automated. Here are how their predictions apply to 2016 US employment statistics. **Black** fields are jobs likely to be automated and white fields are jobs that are likely to remain.

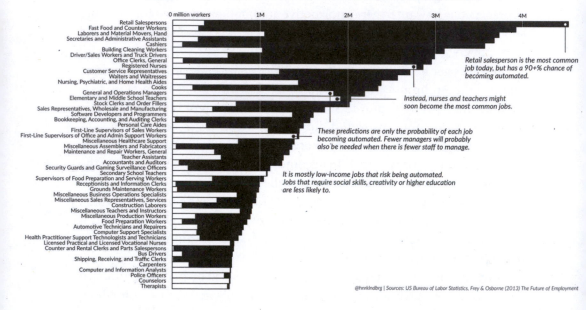

Henrik Lindberg, "The Future of Employment" (2017), Visual Capitalist,
https://www.visualcapitalist.com/visualizing-jobs-lost-automation/. See Chang 2017.

aid and the sharing of technologies, to the victims of capitalism around the world. And it means a total commitment to the ideals of democracy that have been dangerously eroded since the 1970s.

Next question: When will our existing world become better? This is the most difficult question, since no one knows the future. Arguably capitalism is already dead or is at least dying; has exhausted all contradictions, has squeezed nearly every dollar it can out of the natural and human world; and is, finally, cannibalizing the very sociopolitico-environmental world that makes it possible. In short, capitalism is destroying itself. Whether this stunning process of self-immolation creates an "interregnum" than lasts fifteen, twenty-five, or fifty years is impossible to say, but coupled with an environmental crisis that worsens annually and the rising global population, it's very safe to say that our relatively recent ways of being human cannot possibly survive another century. The capitalist subject, the consuming subject, the individualistic and greedy subject of the twentieth century – I'm afraid it's already over. Many of us inside the Green Zones just don't know it yet.

Last question: For whom will this new world of collective consciousness be a *better* world? In some ways this is an easy one. The easy answer, borrowed from Occupy Wall Street protesters, is that it will be better "for the 99%." Less trite: it will be far better for the mass of humanity, the 90 per cent who manage to survive the next few decades (give or take) of environmental catastrophe and societal upheaval. A collectivist mindset that shares instead of exploits, that conserves instead of expends, that preserves instead of pollutes – this would be a radical improvement over the victimization of the masses that passes as our status quo today. As for the 91–98 per cent, many of who live in relative comfort today? That's harder to say. Certainly the joy of conspicuous consumption is nearly over. Those that are wealthy now might not in the near future be able to afford so many cool gadgets, clothes, and imported foods.

Then again, there is reasonable hope that a collectivist future won't be a future of privation. Increased automation isn't just a curse (for workers under capitalism) but a blessing – a point often made by "automation optimists" (Frase 2016, 9). For example, the development of inexpensive printing technologies, from the printing of construction materials to customized human organs, could see the spread of material wealth and bodily health like never before. This is the attraction of techno-futurist scenarios advanced by many, a *Star Trek* future that imagines a marriage of automation and collectivism as strength and abundant wealth – not as just another mechanism to ensure the unequal but largely arbitrary distribution of power and privilege that characterizes life under late capitalism.

An alternative, dystopic future would see unequal access to powerful but inexpensive technologies, a situation that would generate renewed or even heightened inequalities. This would probably look like "rentier class" capitalism or "rentism," an economy organized around copyright, intellectual property, and patents – what Frase calls an "anti-Star Trek world" (2016, 81; Harvey 2014, 251). It follows that collectivists must insist that revolutionary technology be closely controlled and leveraged to make for a far better, instead of a far worse, world.

16 The Anthropocene Condition & the "Original Position"

In the Anthropocene the powerful will still seek ways to reinstate their privilege. They're doing it now. In response to this impulse it's useful to recall, briefly, a famous thought experiment from 1971. In his discussion of the "original position" in *The Theory of Justice*, American liberal philosopher John Rawls asks that we imagine a world remade anew but from behind a "veil of ignorance." Imagine, he says, that we are ignorant of our social standing, gender, intelligence, race, and so on. We might be among the elite few, the lucky, but are more likely to be someone less powerful and wealthy, someone unlucky. In fact, there's a good chance that we would be disadvantaged, for example poor, black, and female. Given a veil of ignorance about our subject position, Rawls asks us to imagine a fair social and political arrangement, one that would do justice to the myriad roles that people play within the resulting social hierarchy. Would we prefer unfettered capitalism or democratic socialism? Despotism or communism? Oligarchy or timocracy? Totalitarianism or liberalism?

In his own thought experiment in the *Republic*, Plato ranks five ideal kinds of constitutions and five kinds of people – souls, characters, or personalities – that would reflect the values of those constitutions. These he lists from the most ideal to the least: the aristocracy of intellectual merit, exemplified by the philosopher's love of wisdom; the timocracy, exemplified by the soldier's love of honour; the oligarchy, exemplified by the merchant's love of money; the democracy, exemplified by the playboy's love

"Peasants for Plutocracy" (2009), Michael Dal Cerro

of freedom; and the tyranny, exemplified by the tyrant's insatiable love of power. For Plato, anyone who understands how to measure the best and worst kinds of constitutions and souls will know, with certainty, that the love of money, freedom, and power are false ideals; that they represent a descending order of decadence, injustice, and unhappiness. Consequently Plato argues in favour of the ideal of wisdom – although he accepts, as a practical matter, that a Spartan-like love of honour is only one step away from the love of wisdom that is best and happiest.

Rawls, fishing in the same pond, asks us to play the part of Plato and think about our world without foreknowledge of our own privilege. If we are likely to be born in a position of relative weakness, then we would probably want a society very different from the one we experience today. Barring arbitrary privilege, most of us would agree to maximize opportunity for equality and justice for all. Hence Rawl's "liberalism."

The Anthropocene condition is a useful concept precisely because it prompts us to rethink our future world in radically different terms, terms that echo (loosely) a Rawlsian "original position." After all, we don't really know how badly we will do in this future; don't know if our individual worlds will be turned upside down by the tragic bad luck of living in the wrong place at the wrong time; don't know if our families, friends, and neighbours will survive the arbitrary impact of Gaia next year or next decade; and don't know if the social, economic, and material infrastructures that currently exist will continue to blunt the impact of destruction and chaos. In fact, we can be pretty sure that these infrastructures won't suffice, since they were never designed for it. And if we don't know how we will fare over the next decades, then perhaps self-interested individuals (even those inclined toward egoism, even liberals) can take seriously the utility and morality of preparing for a world where more of us, not less, will have access to health, wealth, stability, and happiness. Where inequality is understood as a preventable blight, not as an inevitable part of human evolution or innate superiority or even the good luck of being born here and not there, healthy and not sick, white and not brown, male and not female. Where the moral position of collective fairness and wellbeing should trump all the old (almost entirely arbitrary) prejudices and advantages of birth and rank. Where the "we," finally, trumps the "I."

The biggest obstacle is the privilege many of us still enjoy, not as the perilous subjects of anthropocenity but as the privileged subjects of a declining or already dead form of Western capitalism. This scenario is also acknowledged in Plato's *Republic*. In response to Socrates's initial claims about the just life being the best life, that being just, like health, is a "good in itself," Glaucon replies sceptically that justice is actually experienced by people as a burden; that people pretend to be just, publicly praising acts of justice, but privately act quite differently. It's the oldest and most infantile story ever told, usually whispered *sotto voce*: rules are for others, not for me and mine. In this context Glaucon advances a social contract argument, probably the first ever, according to which we only give up our individual advantages (and hence the possibility of egoistic tyranny over others) by compulsion. Why? Because we fear the injustice of other egoists who think just as we do. And so we begrudgingly exchange our own desire for absolute freedom and tyranny over others for collective justice and security.

This calculation still exists in everyday life today – even as pithy philosophies of good government. On the one hand we have the brash American ideal, from the Declaration of Independence, that government exists to enable "life, liberty, and the pursuit of happiness." On the other hand we have the Commonwealth ideal that government exists to ensure the rather more prosaic triumvirate of "peace, order, and good government."

A major shift is coming. If our present moment hinges on the democracy of suffering, then human beings will be obliged, en masse, to reset the conditions of existence. It's tragic that it takes catastrophe to make a better world and to shift human consciousness toward greater empathy. But as Fredric Jameson has famously remarked, people are more willing to imagine the end of the world than the end of capitalism. This is the ultimate meaning (and absurdity) of "capitalist realism" and the dogma of TINA. Ok, then – so be it. If it takes catastrophic climate change and the collapse of capitalism to change the minds of people, then that's what it will be. Of course catastrophe is a telling measure of what is needed to precipitate a more just world – or prepare the world for a time of unparalleled injustice.

17 The Greek Experience

In the 1960s and early 1970s, Marshall McLuhan, Canada's famous media critic, sketched the outlines of a new world with his celebrated phrase "the medium is the message." Human existence, he argued, is not just influenced by radio and television technologies. People also process thoughts differently as a result of new technologies. Followers of McLuhan continued to apply these ideas to our changing culture. Among them, Walter Ong (1982) argued that the contemporary shift in our media-saturated consciousness could be understood by thinking about changes experienced in ancient Greek culture. Socrates and Plato lived during a time of transition from an illiterate oral culture organized around the ear – transmitting tradition and ideas through spoken stories – to a literate culture organized around the eye, around reading, linearity, reason, logic, and observation. The argument is breathtaking in its reductive sweep of history: literacy made possible the birth of Western philosophy, which, in turn, made possible Western science. Or again, the advent of the phonetic alphabet in Plato's Greece – a technology, utterly unique in human history, easy enough for children to master – is the essential bedrock of the Western worldview and its success as a global imperial power.

Ong names the twentieth-century shift in consciousness that McLuhan described as "secondary orality," a return, by analogy, of the educational mediums of speech that dominated the life of Socrates – an illiterate philosopher defined, according to Ong, by his encounter with "primary orality." It's no accident that Plato, on this accounting, wrote his philosophy as a series of Socratic dialogues, as conversations, and in this way bridged the traditional norms of orality with the emergent literacy of his time that he in fact denigrates. And it's no accident that philosophy exhibits an ancient distrust of a writing that distances us from idealized speech, ultimately from the presence (truth, Being) of Socrates. Or, similarly, that philosophy was destined to reject that very tradition in turn, most especially in the 1960s with Derrida's deconstruction, which takes its radical stand, unapologetically, with the actually existing written status of philosophy and thus on the side of contamination, error, and "mere" representation. Finally gone with Derrida is the dream of transcendence – the infantile fantasy, however sublimated – of somehow getting back to the master's authentic speech. In Nietzsche's familiar terms, this god is dead.

The shift in consciousness that accompanies the Anthropocene condition really has no perfect model in recorded history – nothing as inescapably and globally traumatic, "epochal." As Nafeez Ahmed puts it, we must cultivate a "form of collective intelligence that has not yet been fully cultivated and practiced in human history – though it may well have had various past iterations and manifestations. It will be truly 'new,' in the sense of building on the best of what we have so far achieved as a species, to deal with the unprecedented scale and convergence of contradiction and change we now face, together, on the planet" (2018). But the shifts that Ong describes in ancient Athens, like the shift Kant describes as Enlightenment, are arguably instructive. Understood historically, human subjectivity and consciousness is fluid and changing. Massive shifts in material reality, such as that which accompanied literacy and later accompanied industrial capitalism, make for new kinds of subjects or, more loosely, new ways of being human. Marx thought, against Hegel, that economics is the real material condition of reality. And he was right so far as that goes. But the dogma of economics has also blinded us to the real material grounds of the Earth itself and, in turn, blinded us to its far deeper history. Geological history.

INTERGENERATIONAL DEBT

How should we currently value damages to people who will live 500 years in the future? How should we value anthropogenic changes to the biosphere over that period of time? These questions outrun the resources of economics to make sensible evaluations.
(Dale Jamieson, in Gutting and Jamieson 2015)

The environmental epoch called the Holocene no longer functions like an unthought given, an assumption, an eternal constant. This truth we know as well as feel – for it is a troubling feature of everyday public health (see Watts et. al 2018). Who we wish to be in the future will be a function of the ideas we bring to bear and the humanity with which we approach new challenges. For once again the "what" of anthropocenity implies the "who" that comes after the end of our world. Without any doubt at all this "who" will be, or must be, a "we." For only a grouping of human beings, an amassing of individuals, a "mass subject," can save us. The alternative is continued but exacerbated inequality between the elite and the dispossessed, the lucky few and the unlucky many. That is clearly the least wise option of all because the unlucky many will still form, in any case, a collective "we" – but an angry "we" inclined toward violence and retribution. There's nothing more dangerous than a hopeless "we" with nothing to lose. Yet that's the dystopic future our elites are currently fashioning out of a disaster capitalism in free fall.

> As for the economic elite, as the consequences of their own greed and self-interest emerge, they seek, like the Roman oligarchs fleeing the collapse of the Western Empire, only to secure their survival against the indignant mob. (George Monbiot 2018)

18 Unintended Consequences

Let's consider this free fall once again but more concretely, extrapolating worst-case scenarios, since discussions about a shift in human consciousness risk sounding academic, abstract, and painless. Or, in Purdy's view, unhelpful.

The free fall will be most dramatic in the United States. Sprinkle Jesus, automatic weapons, and Washington incompetence over a heaping of mass unemployment, climate disaster, and a failing or compromised infrastructure (roads, bridges, rail, airports, pipelines, satellites, water, sewage, and communications) and you have a deep and abiding disaster. It's easy to predict the broad strokes. At some point over a few short years there will be a confluence of disasters: one or two category-four hurricanes and one or two category-five (or worse) hurricanes along the east coast; tropical storms that bring damaging winds, water spouts, flooding, and tornados along the Gulf of Mexico and further north; permanent coastal flooding as ocean levels rise, most especially in low lying cities like New York; abiding megadroughts and heat waves and, in turn, dozens and probably hundreds of large, mostly uncontrolled, wildfires in the West and Southwest; human-made earthquakes generated in the aftermath of hydraulic "fracking" of shale, most strikingly in states like West Virginia and Michigan where significant seismic activity was once rare; failing pipelines as oil leaks into ground water, tributaries, and lakes, and also onto the streets of suburbia; city water systems contaminated by pollutants including lead, mercury, and e-coli; massive flooding after winter run-off in the Midwest and Northeast;

Tornado, courtesy of the National Oceanic and
Atmospheric Administration photo library

at least a few significant (F4–F5) tornados tearing across the South, Midwest, and Northeast; and riots in cities over limited supplies, insufficient health care, systemic racism, ineffectual leadership, police brutality, and inadequate relief from a Washington underfunded by an increasingly nonexistent tax base. Once rare weather events – like atmospheric "blocking" and "derechos" – will devastate parts of the United States.

> Blocking high: A high pressure area (anticyclone), often aloft, that remains nearly stationary or moves slowly compared to west-to-east motion. It blocks the eastward movement of low pressure areas (cyclones) at its latitude. (Weather Glossary – Terms & Definitions)

> Derecho (noun): A large fast-moving complex of thunderstorms with powerful straight-line winds that cause widespread destruction. (Merriam-Webster)

What you will have, in short, are serial catastrophes across the United States, one after the next, the effects of which will last for years or longer. Throw in one global pandemic, like avian flu or H1N1 or something much worse, coupled with the declining efficacy of traditional antibiotics for regular infections; and, for good measure, throw in one opportunistic cyber attack (from one of many capable enemies) on the power grid, communication system, or finance system. Unless homes and meaningful infrastructure projects are rebuilt and defended after human-made and natural disasters, there will be no real job creation. Recovery will also be impeded by rolling brownouts, blackouts, and worse. Note as well that key stabilizing institutions, like public schooling, will simply collapse. In a shockingly short period of time tens of millions of Americans will die – ultimately reaching one third to one half of the total population. Gang activity and other loose fraternities and alliances, some religious in origin, will compete with police and extort protection services everywhere. In the meantime, military boots-on-the-ground will be stretched thin as they do the Sisyphean work of packing and moving sandbags, delivering supplies, digging trenches and graves, patching roads and bridges, conducting search and rescue, fighting fires, assisting hospitals, feeding the poor, and, moreover, patrolling disaster zones to protect private property and maintain peace and order by arresting looters, squatters, arsonists, rioters, thugs, and murderers. But finally many soldiers won't do any work at all because the federal government won't be able to pay their salaries or keep them housed and fed. Probably they'll return home and, like mercenaries, take up whatever occupation can sustain them given their skill sets.

Just like Private Scranton's experience when he returned to New Jersey from Baghdad, Americans are destined to experience war-like conditions in their own lives, in their own neighbourhoods, over a period of time lasting decades. Red Zones will pop up everywhere and – much worse – won't easily (or ever) revert back to Green. Early on, people with the means will simply withdraw from the ruins – relocating to better regions. Other more desperate Americans will highjack cars, trucks, buses, and planes, or leave on foot, while airlines will simply stop serving entire regions living in poverty and, in any case, living without a working runway. After a passage of time, perhaps as little as ten years, mobility will become less and less possible. And then nearly impossible. Mobility will be a thing of the past, first because gasoline will be hard to get, second because the roads will quickly become impassable and, third, because the locals will in any case impede passage at every turn.

Over the long term it's likely that the United States will break into many regions or self-standing states defended by professional and amateur armed militias who will treat their boundaries like international borders. Executions will be commonplace. Civil war is guaranteed or rather civil wars, since more than one region will fight against another. Local laws will trump federal laws and rights, which will have been rendered meaningless because the United States won't exist anymore. Some regions, for example along the Great Lakes, will remain intact and more or less self-sufficient – unless a big state, like California, expropriates the water, through secure pipelines before the region militarizes. Other regions, like states in the Southwest, will collapse altogether and produce millions of displaced Americans.

Many will die in ignorance, or in full denial, that they are victims of a capitalism rendered traumatically concrete in their own lives as the incapacitating and serial catastrophes of climate change. But no matter.

Riot, courtesy of Pixabay

Cops at riot, courtesy of Pixabay

They will in any case starve, drown, dehydrate, suffocate, bake, and burn or, in the winter months, will freeze; or will be shot, stabbed, beaten, and strangled by someone else impoverished by calamity; or will be crushed when buildings, bridges, and tunnels crumble, either from decades of neglect or from human-made (but at the same time also from natural) earthquakes; or will be swept away, crushed, and buried by landslides that take out roads, bridges, houses, neighbourhoods, and entire towns; or will be mangled in crashes when rail lines and highways are washed out in flash floods and storm surges; or will contract chronic and fatal diseases once controlled by colder and longer winters, better hygiene standards, cleaner water, and access to plentiful drug supplies (and physicians). For many Americans, this hard life will all be neatly rationalized in terms of the Bible – the wrath of God, end times, rapture. True, an even greater number of Americans will understand their plight perfectly well in the light of reason. But they will be utterly stymied by the kind of enduring social divisions and magical thinking that undercuts solidarity and collectivist responses to chaos. The deep tradition of American libertarianism and individualism will not bend to a centralized government or a command economy approach to disaster. Instead America will become a nation of internal refugees, with millions shifting toward the Great Lakes regions and towns along the border with Canada – until the surrounding states impede their mobility. While they can, many Americans will cross the border illegally into Canada, which will still have lots of fresh water and empty land. In the south they'll cross into Mexico. In turn Canada and Mexico will secure their borders against the millions who come. But both countries will fail to stop anyone with even a little determination.

CLIMATE SCENARIOS

At four degrees, the deadly European heat wave of 2003, which killed as many as 2,000 people a day, will be a normal summer. At six, according to an assessment focused only on effects within the US from the National Oceanic and Atmospheric Administration, summer labor of any kind would become impossible in the lower Mississippi Valley, and everybody in the country east of the Rockies would be under more heat stress than anyone, anywhere, in the world today. As Joseph Romm has put it in his authoritative primer *Climate Change: What Everyone Needs to Know*, heat stress in New York City would exceed that of present-day Bahrain, one of the planet's hottest spots, and the temperature in Bahrain "would induce hyperthermia in even sleeping humans." (David Wallace-Wells 2017a)

It's possible, in time, that Canada's population will double, and then double again until it becomes a different country. On the other hand, the slowing of the Gulf Stream in the Atlantic Ocean could mean that Canada, like the northern states, will experience prolonged winters and cool summers. This will destroy the biological diversity of the natural world, which has no chance to adapt, and, in turn, destroy the food supply. So it's possible that Canadian populations will stagnate or decline. If climate and/or desperation allow it, a population surge would help Canadians achieve a relatively high standard of living. Given its massive geography, abundance of fresh water, declining but still arable land, technical know-how, and low population density, Canada could emerge relatively intact in the distant future. On the other hand, a very cold climate could leave it a low-density country of little importance on the world stage – and one that splinters along regional lines. For example, southern British Columbia could easily forge a union with Washington State. In the face of a small population in a weakened Canada and a United States in free-fall, the northern shipping passage would be ceded to Russia. But a middle point between the extremes of prosperity and collapse is perhaps more likely: a Canada run on nuclear power and massive hydroponic projects for food production in urban centres much larger than what currently exist. In almost any scenario Southwestern Ontario will remain a powerhouse, an equal to the powerhouse regions that remain in the United States.

Let's also consider, in even broader strokes, the rest of the world – which will look like better or worse versions of what will befall the United States and Canada. Ultimately Mexico and the countries of Central America may thrive in the face of American refugees. But the old drug cartels, combined with historic and chronic corruption, will fragment Mexico into territorial no-go zones. Assuming they can get past the cartels, American and Mexicans will push further south into Central America, and populations will swell beyond their capacity to feed everyone. The influx of foreigners will cause unorganized and organized pushback, some of it armed. But Americans, better armed yet, will dominate and also push onward into South America. Despite the greater space for development in South America, anti-Americanism is the future of the entire southern hemisphere, where resentment will be common and often deadly. But finally South America will absorb North Americans and be radically transformed. It's possible that countries in South America will merge together or otherwise unify, like the EU, and emerge as a world power in the next century. Probably.

Over the same period, Europe will be mired in refugee crisis after crisis and will face the combustible internal politics that favour fascist solutions to everything. Like everywhere else, food, water, and power will have to be rationed – although good progress in solar technology will lessen the impacts in regions and perhaps entire countries. The elderly, sick, and the poor will continue to die from the heat during summer months and cold in the winter. Major grass and forest fires will be a challenge in Spain, Italy, Germany, and elsewhere; smoke will be a problem. The free flow of citizens within the European Union will slow dramatically, in part because the few hot spots of economic activity (e.g. Germany) won't want them and in part because xenophobia will finally become the unofficial doctrine of the EU. Makeshift refugee camps everywhere, from France to Hungary, will be burned to the ground. Visible minorities will be routinely attacked and armed camps will have to be built to ensure their safety. The locals will insist that the refugees be deported, and the new camps will be subject to constant attacks and the deliberate withholding of food and water. Meanwhile, borders between EU members will be rigorously regulated. Lacking a good economic rationale for the union, and the uselessness of the Euro as a common currency, the EU will cease to function and will collapse. Probably sooner than later. Formerly stable countries will finally splinter along ethnic lines, most producing their own currencies. Like Italy has done for decades, European homes will feature intergenerational family members living on top of each other. Family networks and cooperative alliances, operating like the Mafia, will become the key to survival. In general people won't wander far outside of their own immediate neighbourhoods, and when they do they will

"Fascism is capitalism plus murder." -Upton Sinclair

"Ad Man" (2018), Dwayne Booth

Flood, courtesy of Pixabay

be harassed and threatened with death. The informal black markets will thrive, so much so that a significant portion of international trade (especially in weapons) will be conducted with no oversight at all. The future belongs to whoever controls the seas – and to pirates.

Meanwhile, the Middle East will have long become nearly unlivable, so unbearably hot (temperatures often above 65 degrees Celsius or 150 Fahrenheit with humidity) that all major activity will be reserved for indoors and for evenings. Nations with oil money reserves, like Saudi Arabia, will survive and create massive desalination plants; many countries will adopt solar energy, but most will continue to utilize their own oil reserves. Even so, inequality will reign, with only the rich able to afford air conditioning and the necessities of life. The unlucky poor will, like Syrians today, continue to experience catastrophic droughts, ethnic wars, starvation, sickness, homelessness, unemployment, and a deadly exodus toward Europe and, inevitably, as Europe militarizes its borders, an exodus into the southern half of Africa. The desertification of northern Africa will continue apace, and the old ethnic conflicts will endure everywhere else. Oil wars will be replaced by water wars, so it will be more or less business as usual. Mass murder and genocide will remain common, if not more common. South Africa will defend its border aggressively, or expand it to the north, since it will remain (relatively speaking) a temperate, urbane, and feasible place to live in the Anthropocene. South Africa and the countries of South America will probably establish a significant trade relationship. Shipbuilding, shipping, and the defence of shipping will constitute the foundation of trade in any thriving future.

> "When there is a shortage or scarcity of water, it can be used to make people vulnerable and can be used by combatants, terrorists, or others to put innocents in precarious positions for exploitation, to force migration, and to target vulnerable populations," Goodman said. "You can see that that's happened now in Yemen. You can see the patterns of prolonged drought in Syria, which forced migration."
> (Sherri Goodman in Benson 2018)

To the north, vast Asian and South Asian populations will see their numbers decimated by megadrought, heat waves, starvation, cyclones, tsunamis, typhoons ("hurricanes"), monsoons, diseases, industrial accidents, political inaction, and internal strife over limited resources. Because the world's fisheries will have basically collapsed by mid-century, if not before, millions of people everywhere will be denied access to an easy and essential food source. Some countries could see their populations significantly reduced, perhaps halved from existing levels. The die-off will in any case be tremendous, the disposal of bodies itself becoming a real source of conflict (space, waste, and disease control versus funereal customs). And then war will ignite within and across the nations of Asia. It's likely that conflicts between India and Pakistan will erupt repeatedly. It's also possible that nuclear weapons will be used by either or both countries – unless, just as likely, China or Russia expend efforts to crush their nuclear capabilities for good. While they're at it, they may bomb the nuclear facilities of North Korea.

Meanwhile Australia, like the Middle East, will have become unbearably hot in the summer, with bush fires and suffocating smoke becoming a regular part of life. The sale of breathing masks (particulate respirators) will be a booming part of the economy. And then, in the winter and spring, it will be unbearably wet and flooded. Despite raging wildfires, millions of people will shift south to Tasmania, which in turn will develop hostile political movements to separate from Australia. If possible, hundreds of thousands of Australians will seek access to Commonwealth countries, most especially to Canada. But borders will close before everyone gets a chance.

As for the United Kingdom, the slowing of the Gulf Stream combined with the decline of their fisheries will bring hardship and death. Cold, wet, and miserable, the wealthy will shift to properties in Spain, Portugal, and Italy or to easier climates in the Commonwealth – until those countries shut down their borders. Major conflict between isolated ethnic regions won't be a serious problem because basic survival will make any such campaigns impossible. However, city neighbourhoods will become self-enclosed enclaves defending what little they have, slowly retribalizing the UK and guaranteeing interminable small-scale conflict for decades. The same pattern will hold for Europe more generally, although it's conceivable that the Scandinavian countries to the north will retain the socialist tenor of their fairly homogenous societies. If so, they will provide loose (albeit probably useless) models for other countries for decades to come – and they will likely prosper if they militarize their borders. Still, everything depends on the Gulf Stream.

To the east, Russia will remain as poor as it has been for years, only worse; in a way, this will make Russians more resilient than others accustomed to easy wealth. Old ethnic rivalries among the countries of the Former Soviet Union (FSU), like Georgia and the Ukraine, will erupt into war with neighbours and in some cases civil war. Ethnic genocide will become commonplace and millions will die. Hatreds over racial and religious beliefs will unite groups, but the determining cause will always be conflicts over food, water, energy, and access to the oceans and waterways. This means that Russia, too, will be engaged in multiple conflicts throughout the FSU and, like the United States, will no longer be able to meddle much in international affairs. Finally Russia itself will disintegrate along ethnic lines, and new countries will emerge. Meanwhile, the old plutocrats of the post-Soviet era will still exert tremendous influence over everything, operating like a criminal organization – a mafia but better equipped. Actually governing such a mess may not be in their self-interests. But perhaps they will help to create thriving new cities in the northern parts of the country. Like Canada, they will probably turn to cheap nuclear power and will use their shipping ports to remain influential bullies for a long time.

China, too, will fracture along ethnic lines. But a long history of repression combined with a culture of acquiescence means that China may be able to hold the country together until a renewed spirit of collectivism – in it's case, some form of command economy communism – ensures its survival. China will probably absorb North Korea while it takes over the Asian countries toward the Pacific – with a long-term goal of isolating and annexing Japan. China will survive fundamentally changed but nominally intact after all the social and political chaos at the outset of the Anthropocene condition. It is probably destined to become a stabilizing power in the region and the world's only remaining superpower.

19 Intended Consequences

Just how much suffering is enough to trigger new ways of thinking that can interrupt this free-fall? Or is there really no way of interrupting it?

Certainly many of these futures will happen no matter what we do. That's because some of the dystopic scenarios are already happening; that is, we already occupy this future. Cities in Saudi Arabia and Iran have experienced daytime temperature of 78° C (174° F) and 73° C (165° F) (with humidity). Cities in Spain and Pakistan have experienced heat waves with temperatures above 48° C (120° F) (without humidity) – with death rates at times reaching the thousands. In 2018 Cape Town, South Africa, became the first major city to essentially run out of water. Mountain villages in Sri Lanka, Papua New Guinea, Peru, China, and elsewhere have been buried by landslides. Huge uncontrolled fires are a problem everywhere, from California to Croatia. The New York and Istanbul subways have flooded. In one week in August 2017, Houston experienced record rainfall (19 trillion gallons of rain in about five days) with Hurricane Harvey, rendering 30,000 people homeless, even as Mumbai, India, saw monsoon floods leave millions homeless. These are climate refugees in our own time.

And it'll quickly get worse. Scientists believe that ten highly populated cities will be unlivable in just eighty years (see Bendix 2019). They don't just name places like Shanghai, Beijing, Lagos, New Delhi, and Dubai; they name Miami and Chicago. In fact no place will escape serious consequences. New research indicates that average temperatures in North America cities, for example, will shift hundreds of miles south, even as much as 500 miles (Fitzpatrick and Dunn 2019). So if we don't live to see and feel it, our children almost certainly will.

SUMMER OF 2018

[T]hermometers are bursting in a global heat wave spanning from Japan to Algeria, to Greece, the UK and everywhere in between. This year is on pace to be among the four hottest on record. The other three were 2017, 2016 and 2015. There can be no clearer indication that global warming – the predictable outcome of excessive fossil fuel consumption – is a reality, just as scientists have predicted for decades that it would be.
(Sonali Kolhatkar 2018)

As for the Gulf Stream, it has already slowed measurably – some scientists saying that it's about a century ahead of previous estimates. Just as worrisome, in the summer of 2018 the jet stream (which regulates temperatures in the northern hemisphere) was notably erratic, "meandering like a drunk," and heating up the Arctic (Kormann, 2018). And so on and on. Meanwhile our consumption patterns are still increasing, even in the face of generally slower economic growth since 2008. As a result, it's likely that hundreds of millions of people will die from calamities caused by climate change over the next few decades – starting with the heat, at times literally unsurvivable, and spreading to fire, smoke, drought, landslides, disease, etc.

But that's the environmental future. It doesn't follow that our social, political, economic, and intellectual futures have to be similarly dystopic. Catastrophe doesn't have to mean extinction and the end of civilization. Once again, it should be a wake-up call to do better. If we more aggressively manage our own affairs, if we embrace our agency as responsible subjects, then a full-blown dystopia can be managed and survived with panache and the possibility of living happy, meaningful lives. The future will be challenging, but the story isn't written yet.

So let's start over again and forecast, in the same broad strokes, an alternative future of *intended* consequences that embraces hope and humanity. First of all, the social, cultural, economic, and intellectual divisions in the United States can be managed by political will. The country that elected Donald Trump could just as easily have elected Hillary Clinton or, better, Bernie Sanders. Tomorrow they could easily elect a woman of colour. In any case, presented with competent and compassionate alternatives to the Wall Street consensus, the people will elect a true populist candidate with socialist overtones. And since a strong centralized government is the only government that can possibly address the enormous problems of climate change (and economic free-fall), they will be reelected and remain popular with the people impacted by climate change. Radicalized by reality and by the bumbling incompetence, cynicism, corruption, and contempt of many politicians, the people will continuously agitate in mass demonstrations against the forces of dystopia – which, everyone knows, includes a not insignificant number of Tea Party, libertarian, white supremacist, religious, Republican, and National Rifle Association lunatics. So conflict approaching civil war conditions is probably unavoidable. Fortunately, though, even Republicans will see which way the winds are blowing, and at some point both Democrats and Republicans will fight over better ways to deliver social services and expand the role of government. Taxing corporations heavily will become acceptable once again, not only because corporate leaders will finally understand their own precarious position in American society and agree, however begrudgingly, to pay more. But because Americans in general will be educated about their own taxation history and will better understand that high taxation of obscene profits is actually fair, sensible, and perfectly "American."

Budgets devoted to unwinnable wars that only pad the bottom lines of private contractors will be slashed and applied instead to education, health care, universal basic income, renewable energy, and basic infrastructure. Echoing postwar New Deal society, a much higher percentage of the overall population will work in reconstruction and construction – reconnecting America road by road, solar farm by solar farm. Consequently, when environmental catastrophe strikes the response won't be the craven opportunism of disaster capitalists but the compassion, charity, and collective good will of the majority of Americans. Christianity, incidentally, will thereby regain its meaning and legitimacy in the eyes of an America cynical about the motivations of religious leaders. The church (along with synagogues, mosques, and other places of worship) will in turn become the spiritual and political heart of many struggling communities. Communities and infrastructure will be rebuilt whenever possible, but at times hard decisions will call for retreat to safer locales inland where rebuilding makes more sense. Cities like Miami and New Orleans will have to be abandoned, at least in part.

Instead of breaking into its constituent parts, region by region, and instead of becoming a culture of mutual suspicion and violence, America will remain the United States – more united than before, in fact, not because of wealth and luxury goods but because of shared purpose, a collectivist ethos, compassion for others, and the democracy of suffering. The concrete impact of this shift will be better laws and regulations and the wise implementation of revolutionary technologies. Advances in nanotechnology and biotechnology will revolutionize health; advances in printing technologies will revolutionize the production of pretty much everything; technology devoted to carbon capture will slowly help remediate climate extremes; and targeted automation will render some necessary but unrewarding work obsolete. The gains made will be shared widely. A sizable population of Americans, perhaps as much as two-thirds, will not work in any formal sense but will collect a UBI and do essential volunteer work in their own communities before, during, and after disasters. But they will also take up hobbies and learn new things. Entertainment and culture will thrive. Just like other regions long beset by economic hardship, Americans will discover that happiness requires not just the basic means of survival but also culture, purpose, and a sense of belonging – a community. Happiness is about Yahtzee, not yachts. It's about equality, not equity funds. It has never been about obscene, ostentatious, indefensible wealth for the very few.

And so the world will look a lot more like Denmark and Canada and a lot less like post-Soviet Russia, namely, a world despoiled by plutocrats, mercenaries, and vicious assholes out for nothing but themselves. When nations in distress call for help, the international community will pull together – knowing that they could be next. And when nations, like India, are so desperate that they decide to build new coal power plants, countries like Germany, China, and the United States will launch initiatives, through NGOs and through the United Nations, to help them instead install solar panels on rooftops across the nation. In formerly wealthy households the power will not be sufficient, at least at first. But in formerly impoverished households the addition of lighting and electricity will be utterly transformative. Overall it will be a net gain in happiness, the sharing of a burden made inevitable by the democracy of suffering. The same dynamic holds for the countries of Africa, where the more temperate south will help stabilize and offset the burdens of those living a few hours to the north. Together they will build and manage mega projects to desalinate ocean water and derive energy from solar farms. It's not just cheaper than building fences and garrets, and paying for soldiers and weapons, it's also a lot less dehumanizing for everyone. The same holds for the Middle East, although the switch to solar and wind energy will for years be offset by oil energy. But in a few short decades solar energy and batteries, in particular, will be both cheaper than and just as efficient as oil – at which time the pumping of fossil fuels will end even in the Middle East.

As for the desperate poor, the world community will increase massively its immigration and refugee programs. The model will be Canadian multiculturalism, which has always been the most efficient and generous way of embracing strangers, strangers whose children and grandchildren

Image of Banksy grafitti, London (2009), Anonymous

Wind farm, courtesy of Pixabay

have a way of becoming just as "Canadian" as anyone else, and whose contributions only drive innovation and improve communities. The Commonwealth will absorb many Australians, but most Australians will choose to remain in big cities to the north and south. The Tasmanian population will still explode, the island becoming a favoured destination for travellers everywhere. In fact Hobart could become a significant, thriving world city, more New York than Melbourne. As for the UK, it will still have climate challenges. But populist movements will push for all the social programs that every country will need to preserve peace, order, and good government. There's no mystery here; there's only the dragging of feet in the face of what's obvious.

And that, in a nutshell, is how serial environmental catastrophes will save the Earth from humanity and in turn save humanity from its own worst inclinations. The individualistic rational subject of Enlightenment, turned on its head in the twentieth century, will go down in history as a cautionary tale and the best rationale for creating sustainable communities based on collectivism, equality, freedom, and social justice. Capitalism, itself a manifestation of the Enlightenment subject, will have made this future possible – but in the reverse mirror of its own inverted morality.

In the beginning as in the end, the future still belongs to the people of Earth: To the "we" of a new kind of responsible subject. And that's, of course, because we are together the answer to the question of what is to be done in the Anthropocene condition.

20 This Is the End

The pornography of suffering is probably a permanent counterpart of the democracy of suffering. In time, though, suffering through the environmental and economic apocalypse will be better understood as what it is: a global human tragedy. Fortunately we can live with tragedy in ways that we can't live with apocalypse. If we're careful and thoughtful, creative and smart, the tragedy will provide a bridge to another time and another way of being human. It could be a better time, a better way. So while tragedy isn't nothing, it doesn't forestall a future for humanity. Probably the most pressing condition of anthropocenity is that we face this tragedy now, consciously and with agency – or die trying.

By unpacking the question "What is the Anthropocene condition?" we define and frame the biggest challenge humanity could possibly confront. By thinking philosophically we run the chance, however small, of being truly wise about it. And arguably that isn't nothing, either.

21 Answering the Question: What is the Anthropocene Condition?

From this long reflection we know that the Anthropocene condition, or "anthropocenity," crosses the object and subject registers: what does this condition mean to the earth; what is it doing to human beings, physically and intellectually; and who are we becoming in its wake? The answers to the question are related and sometimes overlapping.

First answer: anthropocenity names our belated recognition of a *condition*, simply put, of life lived in the Anthropocene – the time and space immediately following the Holocene Epoch, which lasted roughly 12,000 years. This condition is necessarily a reflection of the discomfort we feel living outside the environmental niche, or sweet spot, to which human beings are well adapted. Inevitably, the Anthropocene condition orients us toward the unknown future – to what happens next – instead of to the known past.

Second answer: the Anthropocene condition names the intellectual and political effort required to retool the experiences and understandings of civilization to a changing Earth. Although we are obliged to invent new ways, concepts, forms of consciousness, and economies appropriate to our condition, we can't do so in a vacuum. Hence the usefulness of intellectual history, not just as a cautionary tale but as the only available toolkit for thinking about an important aspect of the Anthropocene condition: how did we get here, and how can we change?

Third answer: far from naming a time "beyond" or "after" humanism, the Anthropocene condition names the triumph of old-fashioned humanism. In this respect, our present time completes and radicalizes Kant's own "Copernican Revolution" in thinking. Human beings are at the centre of everything. As such, we *know* the external world by the uncanny impression it makes on us as the emergent subjects of anthropocenity. At present our home *is* and *is not* our home. But it's most definitely the home we have made, or unmade, as "rational" subjects. The Anthropocene condition is the experiential knowledge that comes from the humble recognition that human beings are of and for the Earth, to wit, are *earthlings*.

DAWNING AGE

Yet it is not that the world is becoming entirely technical which is really uncanny. Far more uncanny is our being unprepared for this transformation, our inability to confront meditatively what is really dawning in this age. (Martin Heidegger 1955, 52)

Fourth answer: the Anthropocene condition names a new kind of *subjectivity* forged out of the dialectic of capitalist subjects and actually existing nature in the Anthropocene – both in crisis. Under capitalism, the fortunate enjoyed a peculiar privilege: carefree disregard for the natural environment. The Anthropocene is in this respect a rude awakening, or reawakening, to the perilous conditions of existence in a natural environment not just indifferent to human existence and capitalist economics but increasingly hostile. Human beings have always *fashioned* the world. But capitalist subjects also *transformed* it in totalizing, irreversible ways – making the dreams of humanism concrete, global, inescapable, albeit in the form, once more, of a natural environment to which human beings are no longer well adapted. The subject today is therefore an existential subject but on a species or planetary scale.

Fifth answer: the Anthropocene condition names our *absolute responsibility* for the global conditions of existence in the world today: a responsibility, in short, for the democracy of suffering. Its dominant characteristic is nearly the opposite of Nietzschean-inspired postmodernism, namely, playful irreverence and irresponsibility in the face of the "death of God." And so we come full circle: anthropocenity names the grandiose, debilitating, impossible, yet perfectly justifiable, even realistic, sense of responsibility for the planetary condition created by the Enlightenment project "to know the world." It's on the basis of this feature of anthropocenity that hope exists, most especially concerning the formation of "common ground" among a disparate and apathetic population. That common ground

Bank sign, courtesy of Pixabay

begins with the democracy of suffering after the collapse of the Holocene and the death of capitalism and extends to include the remediation promised by a just, collectivist society for all. And it is sustained by the sort of "green virtues" that Dale Jamieson (2014) enumerates as cooperativeness, mindfulness, simplicity, temperance, and respect for nature. As he says, these virtues "will not solve the problem of climate change on their own but they will help us to live with meaning and grace in the world that we are creating" (in Gutting and Jamieson 2015).

Failure to achieve this burdensome sense of absolute responsibility flirts with a dystopic future, including the extinction of human beings *by the environment*, and the extermination of human beings *by design* – primarily for the convenience of a postcapitalist elite. It should be obvious by now that this dystopia is unravelling as our (mostly) unintended but very deadly present, the time between the easy past and the very hard future. Anthropocenity includes this existential threat to human existence (from *nature in revolt* and from *capitalism ad absurdum*), but also includes the hope, however slim, of a better future based on the transformation of human consciousness in the Anthropocene.

Finally, these five answers to the question of the Anthropocene condition can be reduced to one broad, simple, and perhaps more useful definition. It's just this: the Anthropocene is *the new given* to which we must now submit; "anthropocenity" is *the way we respond to and recalibrate* our own existence against this inevitability. It is a literally a "condition." This formula repackages an old contest between determinism and human freedom, only updated as the most significant problem of our time or indeed of all (human) time.

As for Earth time, geological time, the deep time of Gaia – it is sublimely indifferent to human time and human freedom. For, of course, the Earth is only ours until it isn't.

CONCLUSION

In the time of one human lifespan, more or less coinciding with the American postwar baby boom generation, human beings have produced impacts colossal enough to warrant language, not of events or attitudes, but of *geology*. The dawning panic is consequently moral, ontological, epistemological, and metaphysical. Not paradigm shifting but paradigm busting. Epochal.

Just as Kant, in his own time, contemplated the newfound subject of Enlightenment reason, we too, in the aftermath of Foucault's debased and put-upon subject, must conclude that the Anthropocene condition heralds the birth of a new kind of subject – a new way of being human.

> It's part of our [Indigenous] prophesies … Some of our people have been told that a time will come when these children of the colonizers … [are] going to wake up and see the things that we've always been saying … And they'll form an alliance on the rights of nature.
> (Tom Goldtooth 2018)

This subject of the Anthropocene condition heralds a reinvestment in an anticipatory, aspirational, prophetic philosophy. A philosophy that thinks through the geological metaphor that defines not just our age but an era yet to come. This means reinvesting in old-fashioned beliefs about freedom, democracy, and happiness. It means belief in a better future. It also means that the old-fashioned Marxist ideals of collective freedom and social progress remain the most tantalizing model for thinking the future even if Marxism itself no longer aligns with the postcapitalist realities unfolding as the Anthropocene condition. Most striking of all is how the Marxist commitment to identity forged through labour (species being) has no interpretive foothold at all in a world transformed by automation and the "end of work."

What's fascinating is how unexpected this epochal shift has been. No humanities or social science scholar glimpsed the fact that Enlightenment rationality had prepared the ground for something so radically cut off from the past, something geological. For it wasn't a historian, philosopher, or sociologist but an atmospheric chemist and Dutch Nobel laureate, Paul Crutzen, who in 2002 popularized what we are calling the "Anthropocene" (2002; see also Crutzen and Stoermer 2000). Similarly, two of the most notable figures in the climate wars were once NASA scientists: James Lovelock, the man who coined the Gaia idea, and James Hansen, a trained physicist and prominent climate scientist and activist. We owe quite a lot to whistle-blowing scientists like Hansen. But, it must also be said, we owe almost as much to whistle-blowing journalists, such as Naomi Klein, Paul Masson, David Wallace-Wells, Elizabeth Kolbert, Bill McKibben, and to activist-artists such as Andy Singer, Dwayne Booth, Edward Burtynsky, and Banksy. In a way these two groups have advanced our general understanding of the issues of climate change and political corruption like no others.

SCIENTISTS FOR HUMANITY

If scientists are not allowed to talk about the policy implications of the science, who is going to do that? People with financial interests? (James Hansen 2018)

By the same token, humanistic insight into the changing physical conditions of our future is structurally, intellectually, spiritually, and practically very important. For at last humanity can recognize itself not in an "event," which is far too local a measure, but in a geological disruption – recognize *itself* as the Enlightenment subject that has literally remade the stable physical environment of the Holocene epoch. There is yeoman work to be done unpacking the sociophilosophical significance that the Anthropocene has for life on Earth – but at least it has begun. And in this respect we should also acknowledge those public intellectuals, like Mark Fisher, Kelly Oliver, David Graeber, Roy Scranton, and David Suzuki, who are willing to translate their research in blogs, op-eds, magazines, and newspapers, appear on television, or be interviewed by journalists.

My basic argument is this: the new condition of life in the Anthropocene heralds the dawning of an entirely new kind of recognition, however belated, that originates with but massively inflates Kant's Copernican Revolution in thinking. Kant, recall, understood that there's a veil that lies between us and the noumenal world of the *thing-in-itself* – the independent object world. Kant gave this problem to philosophy – there is an unbridgeable gap between thinking subjects and the objects of the external world – which provided essential direction to Hegel, Marx, and then to everyone else who followed in their wake. Arguably this is no longer the case. For today, lingering in the uncanny interregnum between what was and what is unfolding, the space of our differential consciousness, we *feel* this gap as the condition of everyday life. And yes, we increasingly *know* it, too. Perhaps we will resolve it, as well.

We know that the external world, from the Earth and oceans to the biosphere, is significantly shaped by human activity. True, human beings have always reshaped the Earth by their tireless activity, building pyramids, hydro dams, intercontinental railways, and other great projects. "The bourgeoisie," as Marx and Engels said, "has been the first to show what man's activity can bring about. It has accomplished wonders far surpassing Egyptian pyramids, Roman aqueducts, and Gothic cathedrals; it has conducted expeditions that put in the shade all former Exoduses of nations and crusades" (1848). Subjected to the revolution of steam, electric, and atomic energies (as Ernest Mandel has it), capitalist activity has long transformed the Earth. But that's conscious effort. What's incredible is that even our passive, nonconscious, and often meaningless human activity, our mere existence, has in the last fifty years or so transformed the earth as decisively as any pyramid or hydro dam project did in the past. Or, more truthfully, as decisively as all those projects combined. Even our complacent laziness outstrips the human activity that made it all possible in the first instance. Simple existence – my body, this paper, this keyboard – is complicit in, and reflective of, colossal forces of massive change.

We often measure our individual and collective activities, as in carbon outputs, but routinely fail to ponder how even human inactivity has become so grandly meaningful. We never stop consuming. The meter never stops ticking. At stake is not just the colossal achievements of civilization but the entropic forces of its unravelling.

In 1994 American astronomer and public intellectual Carl Sagan published *Pale Blue Dot: A Vision of the Human Future in Space*. The title originates with a photograph of the same name, taken by the Voyager 1 space probe on Valentine's Day 1990 from a distance of over 6 billion kilometers – close to Neptune. The idea to take the photo was Sagan's, since no one at NASA had thought to turn the camera back at the Earth to see it set against the backdrop of deep space. "It seemed to me," Sagan writes, "that another picture of the Earth [since the Apollo pictures] … might help in the continuing process of revealing to ourselves our true circumstances and condition" (1994, chapter 1).

Instead of a "blue marble" the camera revealed a "pale blue dot" caught within a light ray of the sun. Yet Earth represented as a point of light is really just an intensification of the experiences associated with *Earthrise* and *Blue Marble*. The basic lesson, the one that eluded Arendt and Heidegger, is also the same. As Sagan puts it:

> Our posturings, our imagined self-importance, the delusion that we have some privileged position in the Universe, are challenged by this point of pale light. Our planet is a lonely speck in the great enveloping cosmic dark. In our obscurity, in all this vastness, there is no hint that help will come from elsewhere to save us from ourselves. (chapter 1)

The lesson, existential at its core, is more or less the same one Schopenhauer came up with in 1851 at the end of "On the Sufferings of the World": given that we share the experience of chronic suffering with other human beings, we might learn to indulge one another with more care and kindness. Sagan updates and extends Schopenhauer's conclusion to include kindness to the

Earth itself. As Sagan puts it, the perspective from deep space "underscores our responsibility to deal more kindly with one another, and to preserve and cherish the pale blue dot, the only home we've ever known" (Sagan 1994, 7). That is a beautiful and worthy sentiment.

A technophobic killjoy like Heidegger distrusted technology and distrusted the instrumental, positivistic, nihilistic philosophy that made it all possible. In short, he distrusted Anglo-American analytic philosophy – understood to be a mere handmaiden of uprooted and alienated existence. In this spirit Continental philosophers have long pointed their fingers at analytic philosophy for the horrors of the Holocaust. After all, the machinery of death was enabled by the cold application of reason to a discreet set of "problems"; problems solved by rational bureaucracy, rational census information, rational planning, rational transportation systems, rational engineering, rational waste management, etc. Meanwhile, on the other side of the Atlantic analytic philosophers have long pointed their own fingers at Continental philosophy for the same reason. It's not just that Heidegger worked as a high profile Nazi and never renounced either his own activities or those of his Nazi peers. It's that the Continental penchant for overblown and sometimes mystifying rhetoric, the joyful destruction of meaning, poetic enthusiasm about supermen of the future, phenomenology against empiricism, and interpretation and subjectivity over facts, science, realism, and objectivity – it's that all these tendencies lend themselves so easily to evil perpetrated in the name of vaunted ideals, including dystopic ideals like Aryan supremacy. Death seems to follow wherever, and whichever way, philosophy goes.

We won't solve this impasse, in part because each tradition makes valid points about the other side. But it's increasingly easy to tread a middle path between the extremes. The photographs from space provide an especially good litmus test, since they make it clear that technology doesn't necessarily make us forget the conditions of our human existence, our Being. The truth is, the more mastery and objectivity we have of the Earth, as reflected by the pictures of Earth from space, the more vulnerable and human we feel as subjects. Heidegger simply didn't comprehend any of it – neither McLuhan's world of the 1960s and 1970s nor the birth pangs of our own. Technology, far from distancing us from Earth, has the potential to *collapse* the difference between human beings and the world, subjects and objects. This is just as true for the technologies of reading and writing, once experienced as alienating by the Greeks, as it is for the radio and television of yesterday, the digital technologies of today, and the cheap automated technologies of tomorrow. We have all become increasingly more technologically literate – or, if you prefer, "cyborgs" (Haraway) or "bio-info machines" (Berardi 2011, 23) – the medium of our minds better aligned than ever before with technologies that help us comprehend the Earth and our place on it. Frankly it's nothing to regret or bemoan. It's a regular and necessary feature of what it means to be human.

So sure, technology itself won't *save* us; it's just a tool. But it does *change* us and, if we're wise, it can help us to survive and thrive in the Anthropocene. What will save us are new ways of thinking and existing in the world; new ways of organizing ourselves into larger and more responsive social, political, and economic collectivities; new ways of imagining the good, just, fair, and equitable life; and new ways of living with, instead of against, the Earth. These new ways of being are necessarily charged with meaning, with philosophy – philosophy that still needs to be imagined, charted, and lived. We are the subjects living through this revolution in thinking and living today.

Of course there's more than one kind of revolution. There are intellectual, moral, economic, and social revolutions. There's political revolution. And there's revolution at the level of human consciousness. But the revolution that matters most is the revolution that aligns human existence to the revolutions of the Earth itself. To this end we really don't need to be dragged kicking and screaming, out of Plato's cave, into the light of day, into super-rational being. We really don't need philosophers telling us what to do and who to be. What we really *do* need, though, is for more of us to become philosophically curious about the conditions of existence today – to be (let's just say) *journalist-philosophers* in and of our own perilous time.

In the end wisdom loving isn't just a calling. It's a practice. The Anthropocene has given birth to a condition that makes this practice vitally useful. The final answer to our guiding question is just that: anthropocenity is the call to *be* useful and, by being useful, to *be* responsible. The philosophical life is one that *embodies* wisdom – and then *does something* to show it. It's knowing and showing, thinking and being.

The Anthropocene condition is our chance, maybe our last chance, to matter. We should take it.

Protest sign, courtesy of Pixabay

REFERENCES

350.org. nd. "Climate Science Basics." *350.org*. https://350.org/science/.

Adorno, Theodor, and Max Horkheimer. 2016. *Dialectic of Enlightenment*. London: Verso.

Ahmed, Nafeez. 2018. "Only 'Collective Intelligence' Can Help Us Stave Off an Uninhabitable Planet." *InsurgeIntelligence*, 4 May 2018. https://medium.com/insurge-intelligence/only-collective-intelligence-can-help-us-stave-off-an-uninhabitable-planet-e71916a04a00.

Aitkenhead, Decca. 2008. "James Lovelock: 'Enjoy Life While You Can: In 20 Years Global Warming Will Hit the Fan.'" *The Guardian*, 1 March 2008. https://www.theguardian.com/theguardian/2008/mar/01/scienceofclimatechange.climatechange.

Alexander, Brian. 2017. *Glass House: The 1% and the Shattering of the All-American Town*. New York: St Martin's Press.

Amin, Samir, John Jipson, and P.M. Jitheesh. 2018. "There Is a Structural Crisis of Capitalism." *Monthly Review*, 5 May 2018. https://mronline.org/2018/05/05/there-is-a-structural-crisis-of-capitalism/.

Arendt, Hannah. 1958. *The Human Condition*. Chicago: The University of Chicago Press.

Bacevich, Andrew J. 2008. "He Told Us to Go Shopping. Now the Bill Is Due." *The Washington Post*, 5 October 2008. http://www.washingtonpost.com/wp-dyn/content/article/2008/10/03/AR2008100301977.html.

Bacharach, Jacob. 2018. "Capitalism Is Beyond Saving, and American Is Living Proof." *Truthdig*, 31 August 2018. https://www.truthdig.com/articles/capitalism-is-beyond-saving-and-america-is-living-proof/.

Badiou, Alain. 2018. "Capitalism, The Sole Culprit of the Destructive Exploitation of Nature." *Le Monde*, 26 July 2018. http://theoryleaks.org/text/articles/alain-badiou/capitalism-the-sole-culprit-of-the-destructive-exploitation-of-nature/.

Battiston, Alyssa. 2018. "There's No Time for Gradualism." *Jacobin*, 9 October 2018. https://jacobinmag.com/2018/10/climate-change-united-nations-report-nordhaus-nobel.

Bendix, Aria. 2019. "Scientists Say These 10 Major Cities Could Become Unlivable within Ten Years." *Business Insider*, 11 February 2019. https://www.businessinsider.com/cities-that-could-become-unlivable-by-2100-climate-change-2019-2.

Benjamin, Walter. 1936. "The Work of Art in the Age of Mechanical Reproduction." In *Illuminations*. New York: Schocken Books, 1968.

Berardi, Franco. 2011. *After the Future*. Edinburgh: AK Press.

– 2018. "Bifo: Global Civil War and the Rotting of the White Mind." *Verso Blog*. https://www.versobooks.com/blogs/3689-bifo-global-civil-war-and-the-rotting-of-the-white-mind.

Bloom, Allan D. 1987. *The Closing of the American Mind*. New York: Simon and Schuster.

Bohling, Adam (producer) and Matthew Vaughn (director). 2014. *Kingsman: The Secret Service* [motion picture]. United States: 20th Century Fox.

Borch-Jacobsen, Mikkel. 1994. Unpublished interview with the author.

Brady, Amy. 2018. "The Art and Activism of the Anthropocene, Part I: A Conversation with William T. Vollmann, Chantal Bilodeau, and David Wallace-Wells." *Los Angeles Review of Books*. https://lareviewofbooks.org/article/the-art-and-activism-of-the-anthropocene-part-i-a-conversation-with-william-t-vollmann-chantal-bilodeau-and-david-wallace-wells/.

Brand, Russell. 2013. "Russell Brand on Revolution: 'We No Longer Have the Luxury of Tradition.'" *New Statesman*, 24 October 2013. https://www.newstatesman.com/politics/2013/10/russell-brand-on-revolution.

Briggle, Adam. 2015. *A Field Philosopher's Guide to Fracking: How One Texas Town Stood Up to Big Oil and Gas*. New York: Liveright.

Brynjolfsson, Erik and Andrew McAfee. 2016. *The Second Machine Age: Work, Progress, and Prosperity in a Time of Brilliant Technologies*. New York: W.W. Norton & Company.

Carson, Kevin. 2014. "Peak Capitalism?" *Counterpunch*, 11 August 2014. https://www.counterpunch.org/2014/08/11/peak-capitalism/.

Carter, Jimmy. 1977. "Jimmy Carter: Address to the Nation on Energy," 18 April 1977. http://www.presidency.ucsb.edu/ws/?pid=7369.

Centre for Research on the Epidemiology of Disasters. 2015. *The Human Cost of Natural Disaster: A Global Perspective*. http://www.preventionweb.net/files/42895_cerdthehumancostofdisastersglobalpe.pdf.

Certini, Giacomo and Riccardo Scalenghe. 2011. "Anthropogenic Soils are the Golden Spikes for the Anthropocene." *The Holocene* 21(8): 1269–74.

Chakrabarty, Dipesh. 2009. "The Climate of History: Four Theses." *Critical Inquiry* 35 (Winter): 197–222.

– 2015. "The Human Condition in the Anthropocene." *Tanner Lectures in Human Values*, 18–19 February 2015. https://tannerlectures.utah.edu/Chakrabarty%20 manuscript.pdf.

Chang, Sue. 2017. "This Chart Spells Out in Black and White Just How Many Jobs Will Be Lost to Robots." *Marketwatch*, 2 September 2017. http://www. marketwatch.com/story/this-chart-spells-out-in-black-and-white-just-how-many-jobs-will-be-lost-to-robots-2017-05-31.

Collins, Randall. 2013. "The End of Middle-Class Work: No More Escapes." In *Does Capitalism Have a Future?*, 37–69. Oxford: Oxford University Press.

Cosgrove, D. 1994. "Contested Global Visions: One-World, Whole-Earth, and the Apollo Space Photographs." *Annals of the Association of American Geographers* 84(2): 270–94.

Crutzen, Paul. 2002. "Geology of Mankind." *Nature*. 415: 23.

Crutzen, Paul, and Eugene Stoermer. 2000. "The Anthropocene." *Global Change Newsletter* 41: 17–18. http://www.igbp.net/download/18.31 6f1832132347017758000140l/1376383088452/NL41.pdf.

Doolitte, W. Ford. 1981. "Is Nature Really Motherly?" *The CoEvolution Quarterly* (Spring): 58–62.

Dowd, Maureen. 1999. "Liberties; Trump Shrugged." *The New York Times*, 28 November 1999. https://www.nytimes.com/1999/11/28/opinion/liberties-trump-shrugged.html.

Dufresne, Todd. 2017. "Future." In *Fueling Culture: Energy, History, Politics*, edited by Imre Szeman et al., 170–3. New York: Fordham University Press.

Eaton, George. 2018. "Francis Fukuyama: 'Socialism Ought to Come Back.'" *New Statesman*, 17 October 2018. https://www.newstatesman.com/culture/ observations/2018/10/francis-fukuyama-interview-socialism-ought-come-back.

Fisher, Mark. 2009. *Capitalist Realism: Is There No Alternative?* Winchester, UK: Zero Books.

Fitzpatrick, Matthew C., and Robert Dunn. 2019. *Nature Communications* 10(614). https://www.nature.com/articles/s41467-019-08540-3.

Folley, Aris. 2018. "UN Chief: World Has Less Than 2 Years to Avoid 'Runaway Climate Change.'" *The Hill*, 12 September 2018. http://thehill.com/policy/ energy-environment/406291-un-chief-the-world-has-less-than-2-years-to-avoid-runaway-climate.

Foucault, Michel. 2007. *The Politics of Truth*. Los Angeles: Semiotext(e).

Frank, Thomas. 1997. *The Conquest of Cool: Business Culture, Counterculture, and* the Rise of Hip Consumerism. Chicago: University of Chicago Press.

Frase, Peter. 2016. *Four Futures: Visions of the World after Capitalism*. London: Verso.

Freud, Sigmund. 1895. "Studies on Hysteria." In *The Complete Psychological Works of Sigmund Freud* (Standard Edition), vol. 2, 1–335. Translated by James Strachey. London: Hogarth Press, 1957.

– 1915. "Thoughts for the Times on War and Death." In *The Complete Psychological Works of Sigmund Freud* (Standard Edition), vol. 14, 273–300. Translated by James Strachey. London: Hogarth Press, 1957.

– 1932. "Why War?" In *The Complete Psychological Works of Sigmund Freud* (Standard Edition), vol. 22, 195–215. Translated by James Strachey. London: Hogarth Press, 1957.

– 2012. *The Future of an Illusion*. Edited by Todd Dufresne. Translated by Gregory Richter. Peterborough: Broadview.

Frey, Carl Benedikt, and Michael Osborne. 2013. "The Future of Employment: How Susceptible are Jobs to Computerization?" University of Oxford. http://www.oxfordmartin.ox.ac.uk/downloads/academic/The_Future_of_Employment.pdf.

Fukuyama, Francis. 2012. *The End of History and the Last Man*. London: Penguin.

Gardiner, Stephen M. 2011. *A Perfect Moral Storm: The Ethical Tragedy of Climate Change*. Oxford: Oxford University Press.

– 2016. "Why Climate Change Is an Ethical Problem." *Washington Post*, 9 January 2016. https://www.washingtonpost.com/news/in-theory/wp/2016/01/09/why-climate-change-is-an-ethical-problem/.

Gardiner, Stephen M., and David Weisbach. 2016. *Debating Climate Ethics*. Oxford: Oxford University Press.

Goethe. 1824. *Conversations of Goethe*. Edited by J.P. Eckerman. http://www.hxa.name/books/ecog/Eckermann-ConversationsOfGoethe-1824.html.

Goldtooth, Tom. 2018. "Tom Goldtooth on Climate Change Capitalism." *For the Wild*, 7 June 2018. http://forthewild.world/listen/tom-goldtooth-on-climate-change-capitalism80.

Gould, Stephen Jay. 2007. "Kropotkin Was No Crackpot." *Libcom.org*, 12 February 2007. http://libcom.org/library/kropotkin-was-no-crackpot.

Graeber, David. 2013. "On the Phenomenon of Bullshit Jobs." *Strike Magazine*, 17 August 2013. http://strikemag.org/bullshit-jobs/.

Hansen, James. 2018. "30 Years Later, Former NASA Scientist Wishes He Hadn't Been Right about Climate Change." *CBC News*, 18 June 2018. https://www.cbc.ca/news/technology/james-hansen-global-warming-1.4710713.

Haraway, Donna. 2016. "Tentacular Thinking: Anthropocene, Capitalocene, Chthulucene." *E-flux* 75 (September). http://www.e-flux.com/journal/75/67125/tentacular-thinking-anthropocene-capitalocene-chthulucene/.

– 2016a. "Donna J. Haraway" [Interview]. *Art Forum*, 6 June 2016.
    https://www.artforum.com/words/id=63147.

Harvey, David. 2014. *Seventeen Contradictions and the End of Capitalism*.
    London: Profile Books.

Havelock, Eric. 1963. *Preface to Plato*. London: Harvard University Press.

Hebdige, Dick. 1988. *Hiding in the Light*. London: Routledge.

– 2018. "The Coming Collapse." *Common Dreams*, 21 May 2018.
    https://www.commondreams.org/views/2018/05/21/coming-collapse.

Hedges, Chris. 2018. "Saying Goodbye to Planet Earth." *Truthdig*, 19 August 2018.
    https://www.truthdig.com/articles/saying-goodbye-to-planet-earth/.

Hedges, Chris, and J. Sacco. 2012. *Days of Destruction, Days of Revolt*. New York:
    Nation Books.

Hegel, G.W.F. 1977. *The Phenomenology of Spirit*. Translated by A.V. Miller.
    Oxford: Oxford University Press.

Heidegger, Martin. 1955. "Memorial Address." *Discourse on Thinking*. New York:
    Harper & Row, 1966.

– 1966. "Only a God Can Save Us: The *Spiegel* Interview." *Der Spiegel*, 31 May
    1966. Translated by William J. Richardson. https://archive.org/details/
    MartinHeidegger-DerSpiegelInterviewenglishTranslationonlyAGodCan.

Herrnstein, Richard J., and Charles A. Murray. 1994. *The Bell Curve: Intelligence
    and Class Structure in American Life*. New York: Free Press.

Hickman, Leo. 2008. "Lighten up, Lovelock." *The Guardian*, 5 March 2008. https://
    www.theguardian.com/commentisfree/2008/mar/05/lightenuplovelock.

Holthaus, E. 2015. "The Point of No Return: Climate Change Nightmares Are
    Here." *Rolling Stone*, 5 August 2015. http://www.rollingstone.com/politics/
    news/the-point-of-no-return-climate-change-nightmares-are-already-
    here-20150805.

Husserl, Edmund. 1934. "Foundational Investigations of the Phenomenological
    Origin of the Spatiality of Nature." In *Selected Works*. Translated by Fred
    Kersten. Notre Dame: University of Notre Dame Press, 1981.

Illing, Sean. 2018. "How the Baby Boomers – Not Millenials – Screwed America."
    Vox, 28 April 2018. https://www.vox.com/2017/12/20/16772670/baby-boomers-
    millennials-congress-debt.

Intergovernmental Panel on Climate Change (IPCC). 2014. *Climate Change 2014:
    Synthesis Report. Contribution of Working Groups I, II and III to the Fifth
    Assessment Report of the Intergovernmental Panel on Climate Change*. Edited
    by R.K. Pachauri and L.A. Meyes. Geneva: IPCC.

Jameson, Fredric. 1997. "Five Theses on Actually Existing Marxism." In *Defense
    of History: Marxism and the Postmodern Agenda*, edited by Ellen Woods and
    John Foster. New York: Monthly Review Press.

Jamieson, Dale. 1992. "Ethics, Public Policy, and Global Warming." *Science,*

Technology, and Human Values 17(2): 139–52.

– 2014. *Reason in a Dark Time*. Oxford: Oxford University Press.

Jamieson, Dale, and Gary Gutting. 2015. "What Can We Do About Climate Change?" *The New York Times*, 18 May 2015. https://opinionator.blogs. nytimes.com/2015/05/18/can-green-virtues-help-us-survive-climate-change/.

Kant, Immanuel. 1781/1887. *Critique of Pure Reason*. Edited and translated by Paul Guyer and Allen Wood. Cambridge: Cambridge University Press, 1998.

– 1784a. "Answering the Question: What Is Enlightenment." https:// archive.org/stream/AnswerTheQuestionWhatIsEnlightenment/ KantEnlightmentDanielFidelFerrer2013_djvu.txt.

– 1784b. "Idea for a Universal History from a Cosmopolitan Point of View." http://philosophyproject.org/wp-content/uploads/2013/02/IDEA-OF-A-UNIVERSAL-HISTORY-ON-A-COSMPOLITAN-PLAN.pdf.

Kennedy, Paul. 2017. "Are We F—ked? Decoding the Resistance to Climate Change." *CBS News*, 8 September 2017. http://www.cbc.ca/radio/ideas/are-we-f-ked-decoding-the-resistance-to-climate-change-1.4277614.

Klein, Naomi. 2007. *The Shock Doctrine: The Rise of Disaster Capitalism*. Toronto: Alfred A. Knopf.

– 2015. *This Changes Everything: Capitalism vs. the Climate*. Toronto: Vintage.

– 2018. "Capitalism Killed Our Climate Momentum, Not 'Human Nature.'" *The Intercept*, 3 August 2018. https://theintercept.com/2018/08/03/climate-change-new-york-times-magazine/.

Kojève, Alexandre. 1969. *Introduction to the Reading of Hegel*. Assembled by Raymond Queneau. Edited by Allan Bloom. Translated by James Nichols. New York: Basic Books.

Kolbert, Elizabeth. 2015. *Sixth Extinction: An Unnatural History*. New York: Picador.

Kolhatkar, Sonali. 2018. "As Climate Turns Deadly, Media Are Stuck in Denial." *Truthdig*, 1 August 2018. https://www.truthdig.com/articles/our-climate-has-entered-a-deadly-phase-but-the-media-are-stuck-in-denial/.

Kormann, Carolyn. 2018. "How Climate Change Contributed to This Summer's Wildfires." *The New Yorker*, 1 August 2018. https://www.newyorker.com/science/elements/how-climate-change-contributed-to-this-summers-wildfires.

Lamb, Creig. 2016. "The Talented Mr Robot: The Impact of Automation on Canada's Workforce." *Brookfield Institute*. http:// brookfieldinstitute.ca/wp-content/uploads/2016/07/TheTalentedMrRobotReport.pdf.

Lazier, Benjamin. 2011. "Earthrise; or, The Globalization of the World Picture." *The American Historical Review* 116(3): 602–30.

Leap Manifesto. 2015. https://leapmanifesto.org/whos-on-board/.

Lewis, Simon L., and Mark A. Maslin. 2015. "Defining the Anthropocene."

Nature 515: 171–80.

Lovelock, James E. 1979. *Gaia: A New Look at Life on Earth*. Oxford: Oxford University Press.

– 2006. "The Earth Is About to Catch a Morbid Fever That May Last as Long as 100,000 Years." *The Independent*, 15 January 2006. http://www.independent.co.uk/voices/commentators/james-lovelock-the-earth-is-about-to-catch-a-morbid-fever-that-may-last-as-long-as-100000-years-5336856.html.

Lukacs, Martin. 2014. "New, Privatized African City Heralds Climate Apartheid." *The Guardian*, 21 January 2014. https://www.theguardian.com/environment/true-north/2014/jan/21/new-privatized-african-city-heralds-climate-apartheid.

Lyotard, Jean-François. 1984a. *The Postmodern Condition: A Report on Knowledge*. Translated by Geoff Bennington and Brian Massumi. Minneapolis: Minnesota University Press.

– 1984b. "Answering the Question: What Is Postmodernism?" In *The Postmodern Condition*. Translated by Geoff Bennington and Brian Massumi, 71–82. Minneapolis: Minnesota University Press.

–1984c. "Philosophy and Painting in the Age of Their Experimentation: Contribution to an Idea of Postmodernity." In *The Lyotard Reader*. Edited by Andrew Benjamin, 181–95. Oxford: Basil Blackwell, 1989.

Madondo, Obert. 2015. "RCMP Officer to Bill C-51 Protesters: 'You Could Be Branded a Terrorist.'" *The Canadian Progressive*, 1 June 2015. http://www.canadianprogressiveworld.com/2015/06/01/rcmp-officer-to-bill-c-51-protester-you-could-be-branded-a-terrorist-video/.

Marx, Karl. 1939. *The German Ideology*. Edited by R. Pascal. New York: International.

– 1964. *Economic and Philosophic Manuscripts of 1844*. Edited by Dirk Struik. Translated by Martin Milligan. New York: International.

Marx, Karl and Friedrich Engels. 1848. *The Communist Manifesto*. Translated by Samuel Moore. https://www.marxists.org/archive/marx/works/1848/communist-manifesto/ch01.htm.

Mason, Paul. 2015. *Postcapitalism: A Guide to Our Future*. London: Penguin Random House.

McKibben, B. 2014. "A Call to Arms: An Invitation to Demand Action on Climate Change." *Rolling Stone*, 21 May 2014. http://www.rollingstone.com/politics/news/a-call-to-arms-an-invitation-to-demand-action-on-climate-change-20140521.

McLuhan, Marshall, Quentin Fiore, and Jerome Agel. 1967. *The Medium Is the Message: An Inventory of Effects*. New York: Bantam.

Mitchell, Timothy. 2011. *Carbon Democracy: Political Power in the Age of Oil*. London: Verso.

Monbiot, George. 2016. "No Fracking, Drilling or Digging: It's the Only Way to Save Life on Earth." *The Guardian*, 27 September 2016. https://www.theguardian.com/commentisfree/2016/sep/27/fracking-digging-drilling-paris-agreement-fossil-fuels.

– 2018. "Rebelling Against Extinction." *The Guardian*, 17 October 2018. https://www.monbiot.com/2018/10/19/rebelling-against-extinction.

Mooney, Chris. 2018. "Earth's Atmosphere Just Crossed Another Troubling Climate Change Threshold." *The Washington Post*, 3 May 2018. https://www.washingtonpost.com/news/energy-environment/wp/2018/05/03/earths-atmosphere-just-crossed-another-troubling-climate-change-threshold.

Moore, Malcolm. 2012. "'Mass Suicide' Protest at Apple Manufacturer Foxconn Factory." *The Telegraph*, 11 January 2012. http://www.telegraph.co.uk/news/worldnews/asia/china/9006988/Mass-suicide-protest-at-Apple-manufacturer-Foxconn-factory.html.

Morrison, Toni. 1977. *Song of Solomon*. New York: Vintage.

National Intelligence Council. 2016. *Implications for US National Security of Anticipated Climate Change*. (Rep. No. NIC WP 2016–01). https://www.dni.gov/files/documents/Newsroom/Reports%20and%20Pubs/Implications_for_US_National_Security_of_Anticipated_Climate_Change.pdf.

Nietzsche, Friedrich. 1882. *The Gay Science*. Edited by Bernard Williams. Translated by Josefine Nauckhoff. Cambridge: Cambridge University Press, 2001.

Oaklandsocialist. 2017. "At Our Own Peril." *Oakland Socialist*, 26 July 2017. https://oaklandsocialist.com/2017/07/26/at-our-own-peril/.

Oliver, Kelly. 2015. *Earth & World: Philosophy after the Apollo Missions*. New York: Columbia University Press.

Ong, Walter John. 1982. *Orality and Literacy: The Technologizing of the Word*. London: Routledge, Taylor & Francis.

Ottesen, K.K. 2019. "Environmentalist Bill McKibben: 'Mother Nature Is a Very Powerful Educator.'" *The Washington Post*, 12 February 2019. https://www.washingtonpost.com/lifestyle/magazine/it-was-not-arguing-it-was-fighting-bill-mckibben-on-efforts-to-save-the-planet/2019/02/08/fadad63e-1849-11e9-88fe-f9f77a3bcb6c_story.html.

Peláez, V. 2008. "The Prison Industry in the United States: Big Business or a New Form of Slavery?" El Diario-LaPrensa, New York and Global Research, 10 March 2008. http://www.globalresearch.ca/the-prison-industry-in-the-united-states-big-business-or-a-new-form-of-slavery/8289.

Pendakis, Andrew. 2013. "Franco 'Bifo' Berardi and the Future of Capitalism: 'We Have to Run Along the Line of Catastrophe.'" In *The Economy as Cultural System: Theory, Capitalism, Crisis*. Edited by Todd Dufresne and Clara Sacchetti, 169–76. New York: Bloomsbury.

Petrocultures Research Group. 2016. *After Oil*. Edmonton, Alberta: Petrocultures Research Group.

Plato. 2008. *The Republic*. In *Great Dialogues of Plato*. Translated by W. Rouse. New York: Signet Classics.

Pope Francis. 2015. *Laudato Si': On Care For Our Common Home*. Encyclical Letter. http://w2.vatican.va/content/francesco/en/encyclicals/documents/papa-francesco_20150524_enciclica-laudato-si.html.

Purdy, Jedediah. 2015. *After Nature: A Politics for the Anthropocene*. Cambridge: Harvard University Press.

Rand, Ayn. 2016. *Atlas Shrugged*. New York: New American Library.

Raveendrabose, Shiv. 2018. "Confronting Climate Change Means Confronting Capitalism." *New Socialist*, 11 April 2018. http://newsocialist.org/confronting-climate-change-means-confronting-capitalism/.

Robinson, Kim S. 2015. *Green Earth*. New York: Del Rey.

Romm, Chris. 2018. "Study Confirms Carbon Pollution Has Ended the Era of Stable Climate." *Think Progress*, 1 February 2018. https://thinkprogress.org/carbon-pollution-has-ended-era-of-stable-climate-that-enabled-modern-civilization-fbc0e4a5e476/.

Rorty, Richard. 1984. "Habermas and Lyotard on Postmodernity." *Praxis International*, 4 (1), 33–44. http://www.scienzepostmoderne.org/diversiarticoli/richardrorty-habermaslyotardpostmodernity.pdf.

Ruddiman, William F. 2003. "The Anthropogenic Greenhouse Era Began Thousands of Years Ago." *Climate Change*, 61: 261–93.

Rushton, J.P. 1997. *Race, Evolution, and Behavior*. New Jersey: Transaction.

Sacchetti, Clara. 2013. "Introduction." In *The Economy as Cultural System: Theory, Capitalism, Crisis*. New York: Bloomsbury.

Sagan, Carl. 1994. *Pale Blue Dot: A Vision of the Human Future in Space*. New York: Random House.

Scheidel, Walter. 2017. "The Only Thing, Historically, That's Curbed Inequality: Catastrophe." *Atlantic*, 21 February 2017. https://www.theatlantic.com/business/archive/2017/02/scheidel-great-leveler-inequality-violence/517164/.

Schlosberg, David. 2017. "On the Origins of Environmental Bullshit." *The Conversation*, 25 July 2017. http://theconversation.com/on-the-origins-of-environmental-bullshit-80955.

Schopenhauer, Arthur. 1851. "On the Sufferings of the World." In *Studies in Pessimism. New York: Cosimo, 2007.*

– 1859. *The World as Will and Representation*. Translated by E. Payne. New York: Dover, 1969.

Scranton, Roy. 2013. "Learning How to Die in the Anthropocene." *The New York Times*, 10 November 2013. https://opinionator.blogs.nytimes.com/2013/11/10/learning-how-to-die-in-the-anthropocene.

– 2015. *Learning to Die in the Anthropocene: Reflections on the End of a Civilization*. San Francisco: City Lights.

Shue, Henry. 2018. "Climate Surprises: Risk Transfers, Negative Emissions, and the Pivotal Generation." *Social Science Research Network*. https://papers.ssrn.com/sol3/papers.cfm?abstract_id=3165064.

Smith, Bruce D., and Melinda A. Zeder. 2013. "The Onset of the Anthropocene." *Anthropocene*, December: 1–6.

Smith, Richard. 2016. *Green Capitalism: The God That Failed*. Milton Keynes: College Publications on behalf of the World Economics Association.

Smith, Steven B. 1990. "Hegel and the French Revolution." In *The French Revolution and the Birth of Modernity*. Edited by Ferenc Fehér. Berkeley: University of California Press.

Srnicek, Nick and Alex Williams. 2015. *Inventing the Future: Folk Politics and the Left*. London: Verso.

Steffen, Will, Paul Crutzen, and John R. McNeill. 2007. "The Anthropocene: Are Humans Now Overwhelming the Great Forces of Nature?" *Ambio*. 36 (8): 614–21.

Streeck, Wolfgang. 2016. *How Will Capitalism End? Essays on a Failing System*. New York: Verso.

Suzuki, David. 2013. "Carbon Manifesto." https://www.youtube.com/watch?v=HWPblU8VUyM.

Temin, Peter. 2017. *The Vanishing Middle Class: Prejudice and Power in a Dual Economy*. Cambridge: MIT Press.

Tuhus-Dubrow, Rebecca. 2015. "Impurity: Two Books on the Anthropocene." *Los Angeles Review of Books*, 30 November 2015. https://lareviewofbooks.org/article/impurity-two-books-on-the-anthropocene/.

Union of Concerned Scientists. 1992. "1992 World Scientists' Warning to Humanity." http://www.ucsusa.org/about/1992-world-scientists.html#.WXtjK4jyvIU.

United States Census Bureau. 2015. "International Data Base." http://www.census.gov/population/international/data/idb/worldpopgraph.php.

Varoufakis, Yanis. 2011. "The Road to Bankruptocracy: How Events Since 2009 Have Led to a New Mode of Reproduction." *Personal Website*, 2 March 2011. https://www.yanisvaroufakis.eu/2011/03/02/the-road-to-bankruptocracy-how-events-since-2009-have-led-to-a-new-mode-of-reproduction/.

Wallace-Wells, David. 2017a. "The Uninhabitable Earth." *New York Magazine*, 9 July 2017. http://nymag.com/daily/intelligencer/2017/07/climate-change-earth-too-hot-for-humans.html.

– 2017b. "'The Planet Could Become Ungovernable': Climate Scientist James Hansen on Obama's Environmental Record, Scientific Reticence, and His Climate Lawsuit Against the Federal Government." *New York Magazine*, 12

July 2017. http://nymag.com/daily/intelligencer/2017/07/scientist-jim-hansen-the-planet-could-become-ungovernable.html.

– 2018. "UN Says Climate Genocide Is Coming: It's Actually Worse Than That." *New York Magazine*, 10 October 2018. http://nymag.com/intelligencer/amp/2018/10/un-says-climate-genocide-coming-but-its-worse-than that.html.

– 2019. *The Uninhabitable Earth: Life after Warming*. New York: Tim Duggan Books.

Wang, Jackie. 2018. *Carceral Capitalism*. New York: Semiotext(e).

Watts, Nick, et. al. 2018. "The 2018 Report of the *Lancet* Countdown on Health and Climate Change: Shaping the Health of Nations for Centuries to Come." *The Lancet*, vol. 392, issue 10163, 8 December 2018. https://www.thelancet.com/journals/lancet/article/PIIS0140-6736(18)32594-7/fulltext.

Williams, James. 2000. *Lyotard and the Political*. London: Routledge.

Wolfe, Cary. 2011. *What Is Posthumanism?* Minneapolis: University of Minnesota Press.

Žižek, Slavoj. 2017. "Lessons from the 'Airpocalypse': On China's Smog Problem and the Ecological Crisis." *In These Times*, 10 January 2017. http://inthesetimes.com/article/19787/spaceship-earth-lessons-of-airpocalypse-slavoj-zizek-climate-ecology-smog.

INDEX

origins of, 103–10, 165–8; shifted, 46, 103, 109, 110, 114–15, 118, 148, 155, 156, 165, 166–7, 168, 194

continental (or European) philosophy, xii, xiii, 25, 29, 41, 117, 197

Cosgrove, Denis, 104, 106, 109, 115

Crutzen, Paul, 194

Dal Cerro, Michael, 163

death of God, 38, 39, 40, 41, 110, 123, 166, 190. *See also* Nietzsche, Friedrich

death of philosophy, 6, 29, 38, 40, 42, 122, 166. *See also* postmodernism

democracy of suffering, 66, 71, 118, 134, 148–51, 155–6, 157, 184, 185, 188, 190, 192; described, 71, 118, 148, 150, 156. *See also* globalization of empathy

Derrida, Jacques, 24, 111, 115, 166

Descartes, René, xi, 35, 41

desire, 51, 59, 86, 101, 126, 130, 134, 141, 165; crisis of, 130, 134

Dewey, John, 41

dialectics, xxv, 18–20, 23–4, 25, 27, 43, 56, 58, 123, 156, 190

Donilon, Tom, 79

dystopia (or dystopic), 27, 76, 126, 162, 168, 181, 183, 192, 197; future scenarios as, 127–30, 168–81

earth ethics, 115–16. *See also* Oliver, Kelly

Earthrise (photograph), 103–6, 109–10, 150, 196

education, 7, 13, 17, 25, 72, 121, 148, 150, 158, 159, 184

Einstein, Albert, 17

Ellul, Jacques, 109

empiricism, xi, xii, 19, 42, 197

end of history, 18–22, 72, 112

Enlightenment, xi–xii, xxiv–xxvi, 6–8, 11–17, 24, 25, 27, 30, 34, 35, 36, 38, 40, 42, 43, 44, 46, 49, 72, 110, 113, 115, 116, 117, 139, 167, 188, 190, 193–5, 201; creates opposite values, 25, 27, 30, 34, 35, 38, 40, 43, 44, 72, 113, 195; Foucault on, 11–15, 35, 116; Kant on, xxiv, 6–9, 13, 14–17, 36, 167, 193; Lyotard on, 30, 35, 36, 40; as motto, 7, 11; rendered concrete, 8, 44, 113, 190, 195 (*see also* Anthropocene condition)

essentialism, 39, 122–3, 139

ethics, 34, 54, 56, 61, 115–17, 134, 148, 155

existential, 81–2, 86, 88, 90, 94, 109, 118, 123–5, 190, 192, 196

existentialism, 82, 97; mocked, 123–4

extinction, xxv, 64, 82, 86, 95–7, 100, 126, 147, 154, 183, 192

fascism (or fascist), xvi, 25, 39, 64, 100, 130, 141, 152, 176

fatalism, xvii, 7, 37, 93–5, 148

Fisher, Mark, 54–6, 152, 154, 195

folk politics. *See* politics

Frank, Adam, 81

Frank, Thomas, 66, 71

Frankfurt School, 12, 25, 34, 43, 104, 109

Frase, Peter, xx, xxi, 112, 126, 142, 158, 161–2

Freud, Sigmund (or Freudian), 8, 17, 25, 34, 39, 91, 97, 157

Friedman, Milton, 66, 136

Foucault, Michel, 11–13, 14, 15, 16, 35, 37, 40, 116, 193. *See also* Enlightenment; subject

Fukuyama, Francis, 72

future, xii–xiii, xv–xvi, xx, xxiii–xxvi, 5–8, 11–12, 13–17, 19, 20, 23, 24, 25, 27, 36–7, 39, 40–1, 51, 58, 62, 64, 72, 74, 76, 77, 79, 83, 86, 91, 93, 94, 100,

prediction, xii, xv–xvi, xx, xxiii, 94, 134, 168, 183

prisons, xxiii, 39, 141, 142, 143

public intellectual, 15, 16, 76, 116, 195, 196; philosopher as, ix, 16, 198

Purdy, Jedediah, 9, 103, 123–6, 134, 155–6, 168

Rand, Ayn, 66, 91

rationalism, xi, 18–19, 42, 139

Raveendrabose, Shiv, 75

Rawls, John, 162–4

Reagan, Ronald, 51, 60

recognition (in philosophy), 11, 20, 44, 46, 103, 109, 110, 115–17, 189–90, 195 new kind of, 44, 103, 109, 116–17, 189–90, 195

Red Zones (or no-go), xx, 129, 191, 176. *See also* Klein, Naomi

representation, the philosophy of, 9, 19, 33–4, 43, 106–7, 150, 166

responsibility, 54–6, 68, 110, 114, 118, 123, 134, 154, 190, 192, 197; on absolute, 110, 118, 192

revolution (or revolutionary), xvi, 7–9, 12–17, 25, 33, 36–7, 64, 76, 79, 91, 103, 117, 118, 125, 130, 138, 141–2, 144–5, 151, 162, 184, 189, 195, 198

Rifkin, Jeremy, 116

Robinson, Kim Stanley, 127

Romm, Joseph, 175

Rorty, Richard, 41–2

Rousseau, Jean-Jacques, 7

Rushton, J. Philippe, 66

Rumsfeld, Donald, 66

Sacchetti, Clara, 51

sacrifice zones, xx, 52, 54. *See also* Red Zones

Sagan, Carl, 196–7

Scheidel, Walter, 151

Schlosberg, David, 61, 77

Schopenhauer, Arthur, 9, 97, 98, 123, 196

Scranton, Roy, 79–82, 86, 93, 100, 123–4, 125, 127, 133, 150, 155–7, 171, 195

Sellers, Peter, 109

Sex Pistols, 37

Shue, Henry, 77, 79

Singer, Adam, xiv, 21, 32, 57, 70, 80, 149, 194

Smith, Richard, 74, 76

socialism, xvi, 72, 88, 90, 100, 140, 154, 162

Socrates, 165, 167

species being, 22, 56, 125, 194. *See also* Marx, Karl

Srnicek, Nick, 136–41, 145, 158

Strauss, Leo, 66

Streeck, Wolfgang, 74, 100, 144–6, 158. *See also* capitalism

subject, 7–11, 13, 14, 16, 18–24, 27, 29, 36, 40–6, 49, 55, 58, 79, 86, 88, 103, 106, 107, 109, 112, 113–14, 124, 126, 134, 138–9, 141, 148, 159, 161, 162, 164, 167, 183, 188, 189–90, 193–5, 197–8; in the Anthropocene, xxvi, 9, 11, 46, 79, 86, 88, 103, 109, 134, 139, 159, 164, 167, 193–4, 198; and capitalism, 44, 55–8, 114, 134, 138, 161, 164, 190 (*see also* Capitalocene); Foucault on, 11, 14, 193; Kant on, 6–9, 20, 27, 36, 40, 43, 72, 112, 189, 193, 195; mass subject, 7, 20, 29, 40, 138, 148, 151, 156, 159, 183; as "queer," 13, 35

sublime, 33, 35, 41

Suzuki, David, xxiii, 59, 152–4, 195

Szeman, Imre, 130–3, 136, 144